Deutsche Forschungsgemeinschaft

Biological Exposure Values for Occupational Toxicants and Carcinogens

VCH

Distribution:

VCH, P.O. Box 10 1161, D-69451 Weinheim (Federal Republic of Germany)

Switzerland: VCH, P.O. Box, CH-4020 Basel (Switzerland)

United Kingdom and Ireland: VCH (UK) Ltd., 8 Wellington Court, Cambridge CB1 1HZ (England)

USA and Canada: VCH, 220 East 23rd Street, New York, NY 10010-4606 (USA)

Japan: VCH, Eikow Building, 10-9 Hongo 1-chome, Bunkyo-Ku, Tokyo 113, Japan

ISBN 3-527-27034-5 ISSN 0947-2010

Deutsche
Forschungsgemeinschaft

Biological Exposure Values for Occupational Toxicants and Carcinogens

Critical Data Evaluation
for BAT and EKA Values

Compiled by the Working Group
Setting of Threshold Limit Values in
Biological Material of the Commission
for the Investigation of Health Hazards
of Chemical Compounds in the Work Area
Chairman: Helmut Greim
Head of the Working Group: Gerhard Lehnert

Edited by Helmut Greim and Gerhard Lehnert

Volume 2

VCH

DFG

Prof. Dr. med. Helmut Greim
Senatskommission zur Prüfung
gesundheitsschädlicher Arbeitsstoffe
der Deutschen Forschungsgemeinschaft
GSF Forschungszentrum für Umwelt und Gesundheit
Institut für Pharmakologie und Toxikologie
Ingolstädter Landstraße 1
D-85758 Oberschleißheim

Prof. Dr. Dr. Gerhard Lehnert
Institut und Poliklinik für Arbeits-, Sozial- und Umweltmedizin
der Universität Erlangen-Nürnberg
Schillerstraße 25/29
D-91054 Erlangen

Translator: J. A. Handwerker-Sharman

Die Deutsche Bibliothek – CIP-Einheitsaufnahme

Biological exposure values for occupational toxicants and carcinogens : critical data evaluation for BAT and EKA values / Deutsche Forschungsgemeinschaft. Comp. by the Working Group Setting of Threshold Limit Values in Biological Material of the Commission for the Investigation on Health Hazards of Chemical Compounds in the Work Area. – Weinheim ; Basel ; Cambridge ; New York, NY ; Tokyo : VCH.

NE: Deutsche Forschungsgemeinschaft / Arbeitsgruppe Aufstellung von Grenzwerten in Biologischem Material

Vol. 2. Ed. by H. Greim and G. Lehnert. – 1995
ISBN 3-527-27034-5
NE: Greim, Helmut [Hrsg.]

Composition: Lichtsatz Michael Glaese GmbH, D-69502 Hemsbach
Printing: betz-druck GmbH, D-64291 Darmstadt
Bookbinding: IVB Heppenheim GmbH, D-64646 Heppenheim

Printed in the Federal Republic of Germany

Preface

Biological monitoring has become one of the central and most effective instruments in the prevention of damage to the health after exposure to chemical substances at the workplace. It is an integral part of occupational medical health surveillance and has become indispensable in particular for the monitoring of exposure to substances that can be absorbed by the skin. Biological threshold limit values are necessary for the adequate occupational medical evaluation of the analytical results of biological monitoring. In Germany the Working Group "Setting of Threshold Limit Values in Biological Materials" of the DFG Commission for the Investigation of Health Hazards of Chemical Compounds in the Work Area are involved with this problem. The biological threshold limit values are evaluated on the basis of the latest scientific results and documentation is published in German at intervals in a loose-leaf collection by the VCH Verlagsgesellschaft, Weinheim, Germany. The world-wide interest in biological threshold limit values has prompted the DFG (German Research Foundation) to make the scientific documentation for Biological Tolerance Values for Occupational Exposures and Exposure Equivalents for Carcinogenic Substances (BAT and EKA values) available to interested parties outside German-speaking countries. In 1994 the first volume of Biological Exposure Values for Occupational Toxicants and Carcinogens was published. The occupational medical, toxicological documentation for 19 BAT values and 3 EKA values was presented there. In addition the definition of these threshold limit values, their application and the procedure for their evaluation was described.

I am pleased that one year after the first volume a second set of occupational medical, toxicological documentation for BAT and EKA values can be presented in English. It contains documentation for 12 important BAT values. The particular significance of biological monitoring for occupational exposure to carcinogenic substances is taken into account in 6 EKAs. Important hazardous substances such as arsenic trioxide, ethylene oxide, hydrazine, nickel and vinyl chloride are considered here. In addition to the classic measurement of the concentration of the substance or its metabolites in blood and urine, for the first time measurement of haemoglobin adducts are recommended.

In addition to the documentation for carcinogenic substances with biological exposure equivalents there are 4 further evaluations for acrylonitrile and aromatic amines for which to date no threshold limit values can be formulated. The documentation is intended to help the practising occupational physician in the evaluation of data for these substances gained from biological monitoring. An introductory chapter outlines the importance of the various routes of absorption for hazardous substances in humans. This article is intended to emphasise once again the importance of biological monitoring for exposure to substances that can be absorbed by the skin.

This volume is a tribute to the enthusiastic and highly qualified work of members of the Working Group "Setting of Threshold Limit Values in Biological Materials" of the

DFG Commission for the Investigation of Health Hazards of Chemical Compounds in the Work Area. I wish to thank them for their expertise and personal engagement which is given on a voluntary basis.

I would also like to thank those involved with the secretarial office of the Working Group, in particular the translator of the documentation. Finally I offer my thanks to the Deutsche Forschungsgemeinschaft for its generous support of the activity of the Working Group.

Erlangen, February 1995 G. Lehnert

Foreword

A. Significance and Usage of BAT Values

Definition

The BAT value ("Biologischer Arbeitsstoff-Toleranz-Wert": biological tolerance value for occupational exposures) is defined as the maximum permissible quantity of a chemical substance or its metabolites or the maximum permissible deviation from the norm of biological parameters induced by these substances in exposed humans. The BAT value is established on the basis of currently available scientific data which indicate that these concentrations generally do not affect the health of the employee adversely, even when they are attained regularly under workplace conditions. As with MAK values, BAT values are established on the assumption that persons are exposed at work for at most 8 hours daily and 40 hours weekly. BAT values can be defined as concentrations or rates of formation or excretion (quantity per unit time). BAT values are conceived as ceiling values for healthy individuals. They are generally established for blood and/or urine and take into account the effects of the substances and an appropriate safety margin, being based on occupational medical and toxicological criteria for the prevention of adverse effects on health.

Prerequisites

By definition, BAT values can be established only for such substances which can be taken up by the body in substantial quantities via the lungs and/or other body surfaces (skin, gastro-intestinal tract) during occupational exposure. Another prerequisite for the establishment of a BAT value is that sufficient occupational medical and toxicological data are available for the substance and that these data are supported by observations in man. The data must have been obtained with reliable methods. For the establishment of new BAT values and the annual review of the list, the submission of suggestions and reports of experience with such substances in man is requested.

Documentation

The reasons why a BAT value was established at a particular level are documented by the Commission for the Investigation of Health Hazards of Chemical Compounds in the Work Area in a loose-leaf collection entitled "Arbeitsmedizinisch-toxikologische Begründungen von BAT-Werten". Some of these documents are to appear in English in the present series. They combine a critical review of the available data with comments on the values for the parameters which have been shown in practice to make a useful contribution to occupational hygiene.

Purpose

In the context of specific occupational medical check-ups, BAT values are intended to protect employees from impairment of health at work. They provide a basis for deciding whether the amount of a chemical substance taken up by the organism is harmful or not. When using BAT values, the medical exclusion criteria defined by the "Berufsgenossenschaft" (Employers' Liability Insurance Association) in the guidelines for occupational medical check-ups must be observed. BAT values are not suitable for the derivation, by means of fixed conversion factors, of biological threshold values for long-term non-occupational exposures such as from air pollution or contaminants in food.

Correlations between BAT and MAK Values

When a substance is inhaled under steady state conditions in controlled laboratory experiments, the relationship between the BAT and MAK values can be expressed in terms of pharmacokinetic functions. Under workplace conditions, however, it is not necessarily possible to deduce the level of a substance to which a particular person was exposed from its specific biological parameter in that person because a series of other factors in addition to the amount of substance in the air can determine the extent of exposure of the organism. These factors include the level of physical activity (respiratory minute volume), absorption through the skin and individual variations in metabolic or excretory patterns.

It is therefore particularly difficult to evaluate field studies which describe the relationships between internal and external exposure to substances which can be absorbed percutaneously. Experience has shown that studies of such substances frequently yield discrepant results. These discrepancies are attributed to different levels of dermal exposure under the different study conditions. When evaluating such results to determine relationships between MAK and BAT values, priority should be given to studies in which the data suggest that skin absorption played a minor role.

In addition, the concentrations of substances in the workplace air may vary with time and the biological parameters may not vary to the same extent. Therefore observance of BAT values does not make it unnecessary to monitor the concentrations of substances in the workplace air. This applies especially for local irritants and caustic substances. When evaluating macromolecular adducts of foreign substances the persistance of these adducts should be borne in mind which lead to discrepancies between the external exposure profile and the behaviour of the biological parameters. Similar considerations apply for all strongly cumulative substances such as heavy metals and polyhalogenated hydrocarbons.

In spite of all these interfering factors and the consequent differences in the definitions of MAK and BAT values, the two thresholds are generally based on equivalent effects of substances on the organism. However, for substances for which the MAK value is not established on the basis of systemic effects but because of local irritation, a BAT value can still be based on "critical toxicity" resulting from systemic exposure. In such exceptional cases where the MAK and BAT values are based on different end points, the two values do not necessarily correspond. In this case there is not necessarily any parallelism between MAK and BAT values.

Surveillance

The protection of the health of the individual, which is the reason for establishing BAT values, can be monitored by periodic quantitative determination of the chemical compounds or their metabolites or of biological parameters in biological material. The methods used must be diagnostically specific and sensitive enough for the purpose, acceptable to the employee and practicable for the physician. The sampling time, that is, the measurement strategy, must take into account both the exposure conditions at the workplace and the pharmacokinetics of the substance. As a rule, especially for substances which accumulate in the organism, this may be achieved by taking samples at the end of a working day after an extended period of work (working week). The analytical methods must yield reliable results and meet the requirements of statistical quality control (TRGS 410 of the regulations for hazardous chemicals at the workplace). In the collection "Analysen in biologischem Material" (available in English in the series "Analyses of Hazardous Substances in Biological Materials"), the Commission's "Analytical Chemistry" group has compiled a series of methods which may be considered reliable for this purpose.

Evaluation of Analytical Data

Like any results of laboratory investigations, toxicological analytical data can only be evaluated given knowledge of the whole situation. As well as the other medical findings, especially

- the dynamics of pathophysiological processes
- the short-term effects of exposure-free periods
- the long-term effects of ageing
- the specific conditions at the workplace

must be taken into account.

Results from analyses in biological material are subject to medical discretion. Only the physician who is responsible may interpret the results.

BAT values are established on the basis of the results of scientific studies and practical medical experience.

Allergenic Substances

Depending on individual disposition, allergic reactions can be induced by various kinds of substance, more or less rapidly and in differing degrees of severity after sensitisation of, for example, the skin or respiratory passages. The observance of BAT values cannot provide a guarantee that such reactions will not occur.

Carcinogenic Substances

BAT values are not established for chemical substances which, by their own action or by that of their reactive intermediates or metabolites, are known to cause malignant growths in man or for which there is good evidence of a human cancer risk because at present it is not possible to specify safe levels of such substances in biological materials. Therefore the handling of carcinogens must take place under the conditions described in Section III of the "List of MAK and BAT Values". The determination of carcinogenic substances in biological material is not carried out for the application of a BAT value in the strict sense but rather for the occupational medical detection and quantification of the individual exposure to the compound. Concentrations of the substance or its metabolites in biological material which are higher than those known to correspond to the concentration of the substance in the workplace air are indicative of additional exposure by other routes, usually percutaneous.

For this reason, the Commission investigates the relationships between the concentration of the carcinogen in the workplace air and that of the substance or its metabolites in biological material ("Expositionsäquivalente für krebserzeugende Arbeitsstoffe", EKA: exposure equivalents for carcinogenic substances). From these relationships, the internal exposure which results from uptake of the substance exclusively by inhalation may be determined.

In addition, the considerations expressed in the section "Correlations between MAK and BAT Values" apply to substances which may be absorbed percutaneously.

Mixtures of Substances

BAT values apply as a rule for exposure to pure substances. They are not necessarily applicable for persons exposed to preparations containing more than one toxic substance (blends, mixtures, solutions). This is especially true for BAT values requiring determination of the substance itself or its metabolites. For mixtures of components with similar effects, a BAT value based on a biological parameter can be helpful in the assessment of health risks, as long as it provides a measure of critical clinical-functional effects of the components. The Commission makes every effort to define and publish such criteria for the biological effects of interfering components of mixtures.

B. Occupational Medical Toxicological Documentation of BAT Values

Preliminary remarks

I. In the prevention of damage to health caused by substances at the workplace, the evaluation of biological parameters is becoming increasingly important within the framework of special occupational health surveillance. Medical strategy is based on a principle which has already proved its worth in other fields of application in occupational medicine: the stress-strain (exposure-effect) concept.

Exposure-effect concept

As in other areas of occupational medicine also with toxicological questions the concepts of "Belastung" and "Beanspruchung" corresponding to the English concepts of "stress" and "strain" are used. (In the context of this documentation of BAT values the concepts "Belastung" and "Beanspruchung" have generally been translated by "exposure" and "effect", respectively). Under "Belastung" any influence (dose) is understood which can cause a reaction in the human organism. Changes that are caused by this "Belastung" are, however, described as "Beanspruchung". Individual qualitative and quantitative differences can arise in the strain/effect for a concrete stress/exposure so that the reactions of the organism caused by the stress/exposure receive an individual colouring resulting from the different qualities and capabilities of the person.

The concept can best be illustrated with an example from the area of technology with the rod bending test in the testing of materials. Here it is clear that the bending (strain) of the rod to be tested can only then be predicted from a defined force (stress) working on the rod if the material constant is known. Formulated more generally, this means that the effect on the individual as a function of a given exposure can only be understood if the intervening variables (e.g. functional reserve of an organic system) are known.

The exposure-effect concept is therefore medically of particular importance as it takes into account the congenital or acquired interindividual biological variability and thus places the individual at the centre of attention. In the nature of things, the effect on the person can only be objectified by medical examination.

External Exposure and Internal Exposure

Under workplace conditions the human organism is typically exposed to substances through inhalation. To measure and describe the concentration of the substance in the ambient air of the workplace as the "external exposure" and to limit it by the application of MAK values is an important and indispensable measure of primary prevention. It is, however, much more difficult to quantify other sorts of individual "external exposure" (cutaneous or oral contact with substances at the workplace).

If a substance is absorbed in an inhalational, percutaneous and/or oral way by the body, depending upon its metabolism and kinetics in the organism, a certain concentration of the substance results, the "internal exposure". This quantity, independent of the mode of absorption, is not only a reflection of the total "external exposure"; additionally the intervening variables for the absorption of a substance, such as e.g. the size of the respiratory minute volumes, can influence this quantity. As the "internal exposure" can offer valuable information on the concentration of the substances at the biological site of effect, it is an important parameter of secondary prevention and individual health protection. In many cases the concentration of substances in blood or urine samples can be representative of "internal exposure".

Frequently, however, a biological effect cannot be attributed immediately to the substance but to its metabolites. In these cases the metabolite concentration in biological material must be taken as the parameter of "internal exposure". Thus, the capability of the organism to form and detoxify toxic substances is characterised in such situations as an additional variable intervening between exposure and biological effect.

Effect

The effect resulting from the "internal exposure" can only be deduced from functional changes in the organism. It must, however, be borne in mind that not every deviation of a biochemical parameter from its norm can be immediately regarded as evidence of disease. Changes seem tolerable which even long-term

1. with regular exposure do not result in a disturbance in function or capability to compensate for the effects of exposure,
2. are reversible after the end of exposure,
3. do not intensify the sensitivity of the organism to other external influences, in particular, physical, chemical and microbiological ones,
4. do not endanger the progeny.

II. The Commission wishes to reveal with its "arbeitsmedizinisch-toxikologische Begründung von BAT-Werten" ("Occupational Medical Toxicological Documentation of BAT Values") which observations, experiences and considerations were decisive for the establishment of a BAT value taking the exposure-effect concept as a basis. At the same time it wishes to highlight with the documentation gaps in the data and provide impulses for further research.

The treatise on each substance is subdivided into several main sections.

Front page

Chemical substance
BAT value/EKA correlation
Sampling time
Date of evaluation
Synonyms
Formula (sums and structure)
Molecular weight
Melting point
Vapour pressure at 20 °C
Solubility in water
MAK/TRK [last established]

As an *introduction* a few additional chemical-physical constants which are relevant to exposure and metabolism may be taken into consideration.

Physical state of the substance at the workplace: [gas, vapour, aerosol, mist or dust] Additional exposure to isomers, contaminants etc.

Information on whether the substance is still used industrially or for how long it has been used or produced.

1. Pharmacokinetics

1.1 Absorption

Inhalational, dermal, gastrointestinal – main route of absorption

Inhalational: Pulmonary absorption rate during, at the end of and after the shift. Parameters which influence absorption: e.g. solubility or particle size, physical activity, body weight.

Dermal: Information on the rate of absorption [$mg/cm^2/h$], determined in vivo/in vitro. What role does dermal absorption play in the total absorption?

Gastrointestinal: Information on the rate of absorption and the possibility of occupational and non-occupational oral exposure (contamination of the hands, clothing, food, drinks).

1.2 Elimination

Role of metabolism and storage in elimination; extent and speed, routes of excretion (amount in expired air, urine, faeces).

Elimination half-life of the substance and important metabolites. Signs of accumulation of the substance or of metabolites during one day, a week or years.

1.3 Metabolism

Diagram of metabolism in which the main metabolic pathways as well as proven and postulated metabolites are shown.

Information on the dose dependency of metabolism. At which concentrations are signs of saturation and thus a deviation from first order kinetics to be expected?

Information on known interactions, e.g. with alcohol, drugs, food components.

- Pharmacokinetics (half-life)
- Liability to interference (exogenous and endogenous factors)
- Influence of illness and pregnancy

2. Critical Toxicity

In this section it is explained which toxic effects were decisive for the establishment of the BAT value, whereby often a cross-reference is possible to the loose-leaf collection "Toxikologisch-arbeitsmedizinische Begründung von MAK-Werten" (in English translation in the series "Occupational Toxicants", obtainable from VCH Verlagsgesellschaft mbH, Weinheim, FRG).

3. Exposure and Effect

In a summary of the relevant literature it is shown which relationships exist between external and internal exposure as well as between internal exposure and effect.

3.1 Relationship Between External and Internal Exposure

Short description of the relevant data from

a) laboratory studies on volunteers
b) field studies on occupationally-exposed persons
c) simulation studies – pharmacokinetic modelling

3.2 Relationship Between Internal Exposure and Effects

Short description of the relevant data for the internal exposure and sensitive and specific reactions (see critical toxicity) from:

a) laboratory studies on volunteers
b) field studies on occupationally-exposed persons
c) simulation studies – pharmacokinetic modelling

4. Selection of the Indicators

The increase in individual safety which BAT values are intended to achieve can only be attained if the methods used for preventive examinations are characterised by as high a diagnostic specificity and sensitivity as possible. At the same time the sampling method should be acceptable for the employee and practicable for the attending physician. These criteria and the available knowledge of critical toxicity, as well as the relationship between exposure and effect are taken into account in the selection of the biological indicators. These can be either parameters of internal exposure or of effect. Blood and urine are the preferred examination materials; in certain cases also exhaled air, secretions, body appendages and biopsy material can be considered. The selection of the indicators is substance-specific.

a) Description of all parameters which come into question

Advantages and disadvantages (specificity, measurability, practicability, influence of exogenous and endogenous parameters)

b) Selected parameter(s)

Information on kinetics (immediate increase after exposure, increase in concentration during exposure, decrease or further increase after the end of exposure, decrease in the concentration of the parameter after the end of exposure, type of decay curve and biological half-life, accumulation during the working week, months or years, main storage organs).

5. Methodology

Is the use of particular methods of analysis required for monitoring internal exposure using the suggested BAT/EKA value?

Which methods are suggested by the working group "Analyses of Hazardous Substances in Biological Materials"?

Which methods of analysis have proven reliable and meet the standards of statistical quality control?

Recommendations for sampling, transport and storage. Indication of the possibility of contamination.

6. Background Exposures

Information on the reference values ("normal values") found with the suggested methods. Indication of possible sources of background exposure from the environment or lifestyle (e.g. food, smoking, drugs, implantations).

7. Evaluation of BAT/EKA Values

Summary for the selection of the exposure/effect parameter(s) and the setting of sampling times and the value. Medical evaluation of the data according to clinical experience.

8. Interpretation of the Data

A summary of the clinical-medical experience available so far should simplify the difficult task of assessing the examination data medically from the overall clinical situation.

Information on the importance of adhering to the sampling time

Indication of possible errors in sampling, storage and transport

Indication of analytical problems (use of a particular method, necessity of hydrolysis)

Information on the possibility of an influence on the values measured by the exposure (percutaneous absorption, non-occupational exposure, interindividual and intra-individual fluctuations due to endogenous or other background levels, co-exposure, significant influence of smoking, drugs, physical activity)

Indication of influences due to the individual (influence of illness, pregnancy)

9. References

Important statements should be attested by accessible quotations from the original literature. There is no claim to completeness, suitable reviews are permissible.

Contents of Volume 2

Documentation for Carcinogenic Substances

With Biological Exposure Equivalents

Without Biological Exposure Equivalents

List of Authors and Date of Compilation

Texts	Authors	Compilation
Acetylcholinesterase Inhibitors	J. Lewalter	May 1985
Acrylonitrile	H. M. Bolt	
	J. Lewalter	May 1993
4-Aminodiphenyl	J. Lewalter	November 1987
Aniline	J. Lewalter	May 1985
Addendum	J. Lewalter	May 1993
Arsenic trioxide	K. H. Schaller	May 1993
2-Bromo-2-chloro-1,1,1-trifluoroethane (Halothane)	D. Henschler	June 1982
2-Butanone	J. Angerer	March 1988
4,4′-Diaminodiphenylmethane	J. Lewalter	May 1993
N,N-Dimethylformamide	K. H Schaller	
	H. Drexler	May 1993
Ethylene	H. M. Bolt	May 1993
Ethylene oxide	K. Norpoth	May 1994
	H. M. Bolt	
Hexachlorobenzene	E. Schumacher-Wittkopf	
	G. Lehnert	February 1984
γ-1,2,3,4,5,6-Hexachlorocyclohexane	N. Wichmann	
	B. Eistel	September 1988
Hydrazine	J. Lewalter	May 1991
2-Naphthylamine	J. Lewalter	December 1991
Nickel and sparingly soluble nickel compounds	J. Angerer	March 1990
Nitrobenzene	J. Lewalter	February 1986
Parathion	J. Lewalter	May 1985
Tetraethyllead	H. M. Bolt	May 1994
Tetrahydrofuran	J. Lewalter	May 1994
Vinyl chloride	G. Müller,	
	K. Norpoth	February 1986
Xylenes	J. Angerer	February 1984

List of Substances and Synonyms

Acetylcholinesterase Inhibitors

Acrylonitrile*

Alkyl phosphates .... see Acetylcholinesterase Inhibitors

Alkyl sulphates .... see Acetylcholinesterase Inhibitors

4-Aminodiphenyl*

p-Aminodiphenyl .... see 4-Aminodiphenyl

Aminobenzene .... see Aniline

2-Aminonaphthalene .... see 2-Naphthylamine

Aniline*

Arsenic trioxide*

Benzene-γ-hexachloride .... see γ-1,2,3,4,5,6-Hexachlorocyclohexane

Butylene oxide .... see Tetrahydrofuran

2-Bromo-2-chloro-1,1,1-trifluoroethane (Halothane)

2-Butanone

Carbamates .... see Acetylcholinesterase Inhibitors

Chloroethylene .... see Vinyl chloride

Cyanoethylene .... see Acrylonitrile

Diamide .... see Hydrazine

4,4′-Diaminodiphenylmethane*

p,p′-Diaminodiphenylmethane .... see 4,4′-Diaminodiphenylmethane

Diethylene oxide .... see Tetrahydrofuran

***N,N*-Dimethylformamide**

Dimethylbenzene .... see Xylenes

DMF .... see *N,N*-Dimethylformamide

E 605 .... see Parathion

Ethene .... see Ethylene

* carcinogenic substance listed in Section III of the List of MAK and BAT Values

Ethylene*

Ethylene oxide*

Ethyl methyl ketone .... see 2-Butanone

N-Formyldimethylamine .... see *N,N*-Dimethylformamide

Halothane .... see 2-Bromo-2-chloro-1,1,1-trifluoroethane

γ-HCH .... see γ-1,2,3,4,5,6-Hexachlorocyclohexane

Hexachlorobenzene

γ-1,2,3,4,5,6-Hexachlorocyclohexane

Hydrazine*

Lead tetraethyl .... see Tetraethyllead

Lindane .... see γ-1,2,3,4,5,6-Hexachlorocyclohexane

MDA .... see 4,4′-Diaminodiphenylmethane

MEK .... see 2-Butanone

4,4′-Methylenedianiline .... see 4,4′-Diaminodiphenylmethane

Methyl ethyl ketone .... see 2-Butanone

2-Naphthylamine*

β-Naphthylamine .... see 2-Naphthylamine

Nickel* and sparingly soluble nickel compounds

Nitrobenzene

Nitrostigmine .... see Parathion

Organic phosphates .... see Acetylcholinesterase Inhibitors

Organic sulphates .... see Acetylcholinesterase Inhibitors

Oxolane .... see Tetrahydrofuran

Paraphos .... see Parathion

Parathion

Perchlorobenzene .... see Hexachlorobenzene

Phenylamine .... see Aniline

---

* carcinogenic substance listed in Section III of the List of MAK and BAT Values

* carcinogenic substance listed in Section III of the List of MAK and BAT Values

Working Group "Aufstellung von Grenzwerten in biologischem Material" (Setting of Threshold Limit Values in Biological Material)

Head: Prof. Dr. med. Dr. h. c. G. Lehnert
Institut und Poliklinik für Arbeits-, Sozial- und Umweltmedizin der Universität Erlangen-Nürnberg
Schillerstr. 25/29
D-91054 Erlangen

Members of the Working Group:

Prof. Dr. rer. nat. J. Angerer
Institut und Poliklinik für Arbeits-, Sozial- und Umweltmedizin der Universität Erlangen-Nürnberg
Schillerstr. 25/29
D-91054 Erlangen

Prof. Dr. med. Dr. rer. nat. H. M. Bolt
Institut für Arbeitsphysiologie
Ardeystraße 67
D-44139 Dortmund 1

Prof. Dr. med. H. Greim
GSF Forschungszentrum für Umwelt und Gesundheit
Institut für Pharmakologie und Toxikologie
Ingolstädter Landstraße 1
D-85758 Oberschleißheim

Dr. rer. nat. J. Lewalter
Ärztliche Abteilung, Arbeitsstoffanalytik
Bayer AG
D-51368 Leverkusen

Prof. Dr. med. K. Norpoth
Institut für Hygiene- und Arbeitsmedizin
Universitätsklinikum der Gesamthochschule Essen
Hufelandstraße 55
D-45147 Essen

Prof. Dr. med. H. W. Rüdiger
Universität Wien
Abteilung Arbeitsmedizin
Med. Klinik IV
Währinger Gürtel 18–20
A-1090 Wien

Dipl.-Ing. K. H. Schaller
Institut und Poliklinik für Arbeits-, Sozial- und
Umweltmedizin der Universität Erlangen-Nürnberg
Schillerstr. 25/29
D-91054 Erlangen

Scientific Advisors:

Prof. Dr. med. Dipl. Chem. G. Triebig
Institiut für Arbeits- und Sozialmedizin
der Universität Heidelberg
Hospitalstraße 1
D-69115 Heidelberg

Prof. Dr. med. H. Valentin
Institut und Poliklinik für Arbeits-, Sozial- und
Umweltmedizin der Universität Erlangen-Nürnberg
Schillerstr. 25/29
D-91054 Erlangen

Secretary:

c/o Institut und Poliklinik für Arbeits-, Sozial- und
Umweltmedizin der Universität Erlangen-Nürnberg
Schillerstr. 25/29
D-91054 Erlangen

Introduction to Biological Monitoring

Determination of individual uptake of chemical substances at the workplace

In recent years biological monitoring has become an integral part of occupational medical health care. The determination of hazardous substances or their metabolites in biological materials has often revealed internal exposures which were unexpectedly high in view of the very low ambient air values; such exposures could, for example, be a result of underestimation of the ability of the substances to penetrate the skin. In the following the advantages of biological monitoring for the determination of the absorption of hazardous substances by an individual are presented.

Within the limits set by metabolic and toxicokinetic parameters, under experimental conditions in the laboratory when the substance is taken up entirely by inhalation, there is a relationship between the threshold limit values of external exposure (Maximale Arbeitsplatzkonzentration – Maximum Concentrations at the Workplace = MAK or Technische Richtkonzentration für krebserzeugende Arbeitsstoffe – Technical Exposure Limits for carcinogenic substances = TRK) and those of internal exposure (BAT and EKA values) (2). These conditions, however, do not adequately reflect the exposure variations associated with industrial processes nor interindividual differences in exposure. Therefore the estimation of the actual individual exposure to the hazardous substance is only possible through biological monitoring. As the internal exposure is determined to a great extent by individual differences in *absorption, distribution, metabolism* and *excretion* of a substance as well as by the *route of absorption,* the adherence to BAT values does not mean that also the MAK values are adhered to or vice versa, that with the exceeding of the BAT values the MAK values are also exceeded.

While ambient air investigations are used particularly for the monitoring of collectives and can also reveal short-term exposure peaks, analyses in biological material should identify an individual danger and recognise cumulative exposure. Worries on the part of the employees about surveillance through biological monitoring to the effect that increased values would be equated with personal negligence or decreased resistance and could lead to negative sanctions must be taken very seriously particularly in times of economic recession. Unlike the results of the ambient air investigations at the workplace the biological monitoring values are findings which are subject to medical discretion and therefore are not allowed to be made known to the employer. Medical discretion, which is also the basis of every other occupational medical activity, is thus the guarantee against misuse of the results of biological monitoring by third parties.

In the following the individual as well as the general differences in the uptake and absorption of hazardous substances are presented. The main routes of absorption are by inhalation and through the skin. The attempt has been made in table 1 to emphasise the most important routes of absorption for the substances in the current BAT/EKA list and to provide a quantitative evaluation of the absorption. In addition attention is drawn to the designation "cutaneous absorption" (H) for these substances as in the MAK list. The mostly accidental ingestion of hazardous substances will, due to its minor occupational medical relevance, not be referred to.

Tab 1: Routes of absorption of substances with BAT and EKA values according to data in the occupational medical toxicological documentation of Biological Tolerance Values for Occupational Exposures (BAT values) and Biological Exposure Indices (BEI)

Substance (BAT)	Notation (TRGS 900)	Inhalation	Skin absorption	Ingestion (non-occupat.)
Acetone		+++		
Acetylcholine esterase inhibitors	H	+	+++	
Aluminium		+++		yes
Aniline	H	+	+++	
2-Butanone		yes	yes	
p-tert-Butyphenol	H	++	yes	
Carbon disulphide	H	yes	yes	
Carbon monoxide		+++		
Chlorobenzene		?	?	
Dichloromethane		+++		
Dimethylformamide	H	++	+	
Ethoxyethanol	H	+	+++	
Ethoxyethyl acetate	H	+	+++	
Fluorine: hydrogen fluoride	H	+++		
Halothane		+++		
Hexachlorobenzene		yes	yes	
n-Hexane		+++		
2-Hexanone		+++		
Lead		+++		yes
Lindane	H	yes	yes	
Mercury, inorg. comps		+++		yes
Mercury, org. comps	H	yes	yes	yes
Methanol	H	yes	yes	
Methyl isobutyl ketone		+++		
Nitrobenzene	H	++	+	
Parathion	H	+	++	
Phenol	H	+	++	
Styrene		++	yes	
Tetrachloroethene		+++		
Tetrachloromethane	H	yes	yes	
Toluene		++	yes	
1,1,1-Trichloroethane		++	yes	
Trichloroethene		++	yes	
Xylenes	H	++	yes	

Substance (EKA)	Notation (TRGS 900)	Inhalation	Skin absorption	Ingestion (non-occupat.)
Alkali chromates		+++		
Arsenic		++	yes	yes
Benzene	H	++	yes	
Cadmium		+++		yes
Cobalt		+++		yes
Ethylene oxide	H	++	yes	
Hydrazine	H	yes	yes	
Nickel		+++		yes
Pentachlorophenol	H	yes	yes	yes
Vinyl chloride		+++		

+++ = most important route of absorption at the workplace (>90 %)
++ = important main route of absorption at the workplace (>50 %)
+ = relevant route of absorption at the workplace (<50 %)
yes = no quantitative data, but absorption by this route has been described

1 Inhalation

1.1 Physiology

The alveoli present an absorption surface with an area of 50 to 90 m^2. Metabolism in the lung takes place in the alveolar membrane which is less than 1 µm thick, a lipid membrane with a "large number of pores". It is permeable for hydrophilic and lipophilic substances as well as for large molecules with a molecular mass up to 5000 d (6). Diffusion is dependent on the concentration difference between alveolar air and blood and the alveolar air concentration is determined by the respiratory minute volume as metabolism in the lung takes place very rapidly. An adult ventilates at rest approx. 7.5 l air per minute. During physical stress the greater tidal volume and increased respiratory rate lead to an increase in the respiratory minute volume to more than 100 l.

1.2 Inhalation of hazardous substances

The concentrations of a hazardous substance measured in the air at the workplace are generally not a sufficiently exact measure of the internal exposure of an employee which they cause. Often a stationary ambient air measurement is carried out on the assumption of an even distribution of the substance in the form of gases, fumes or suspended matter in the room. For the individual employee, however, changing activity or time spent in various work areas often results in greatly differing concentrations of the substance in the respiratory air. The comparison with personal measuring procedures (actively with a suction pump or passively through monitors) often reveals marked differences from the results of stationary measurements.

However, even personal air sampling cannot describe exactly the amount of substance inhaled. An increased respiratory minute volume of exposed individuals, for example as the result of heavy physical work, constitutionally low muscle strength, poor fitness or pathological cardiopulmonary conditions, leads to relatively higher amounts of inhaled substances. The absorption behaviour is also different with deeper breathing than with more shallow breathing. So far the changes in absorption behaviour of hazardous substances during work at higher atmospheric pressure have hardly been researched. Furthermore, with inhalation of suspended matter and aerosols the absorption is also determined by the efficacy of the mucociliary cleansing mechanism. The longer persistence in the tracheobronchial tree of particle-bound substances due to decreased ciliary activity (among other things due to the inhalational consumption of tobacco) leads to a longer exposure period. On the other hand by coughing up and swallowing hazardous substances bound on dust a significant gastrointestinal uptake can occur.

From table 1 it can be seen that for most of the hazardous substances of the BAT/EKA list inhalation is the main route of absorption. 17 of the 44 substances are absorbed almost exclusively via inhalation and 9 mainly by inhalation.

2 Skin absorption (percutaneous absorption)

2.1 General remarks

At present 139 substances and thus more than 1/4 of all of the hazardous substances listed in TRGS 900 (Technical Guidance Note for Hazardous Chemicals) or in addendum II of the Hazardous Substances Ordinance are regarded as absorbable via the skin ("H" notation). In general substances with both lipophilic and hydrophilic parts have the best skin penetration (5, 23). However, the prerequisites for the "H" notation have so far not been strictly defined in the German Commission for the Investigation of Health Hazards of Chemical Compounds in the Work Area. In other countries work is in progress to establish criteria for classification of substances according to their ability to penetrate the skin (sk = skin notation, H = Haut; Huid) (7, 17). In particular members of the American BEI (Biological Exposure Indices) Committee have been concerned with this problem (5). In a recent study it is suggested that all hazardous substances with a dermal LD_{50} of less than 1000 mg/kg should be regarded as absorbable through the skin until the occupational medical relevance of this exposure route has been clarified definitively (12).

As can be seen from table 1 with at least 6 of the substances listed the transdermal route of absorption must be regarded as the more important. With direct skin contact this should apply for further substances. However, it should be noted that the quantitative statements on the contribution of the various routes of absorption are mostly only based on a few investigations of small collectives and for many substances are not available at all so that the possibility of skin absorption as a rule has only been demonstrated qualitatively.

TRGS 150 (Technical Guidance Note for Hazardous Chemicals) "Immediate Skin Contact" explicitly prescribes biological monitoring for the quantification of exposure for substances with BAT values or EKAs. Skin contact should also be taken into account in particular during occupational medical health surveillance according to the berufsgenossenschaftliche Grundsätzen (Guidelines issued by the Employers' Liability Insurance Association) G3, G4, G5, G6, G8, G9, G10, G13, G18, G29, G33, G40 (contact with substances absorbable through the skin).

2.2 Basic skin physiology and toxicokinetics

The human skin is made up of the vessel-free epidermis (top layer) with the stratum corneum responsible for the barrier function (horny layer), the corium rich in vessels and connective tissue (dermis, true skin) which regulates the supply of blood to the skin and in which the uptake of foreign substances into the blood stream takes place and the subcutis (subcutaneous adipose tissue) which has mainly physical functions (padding, heat insulation) and represents the most important energy depot of the human organism (lipopexia) (Fig. 1). The thickness of the subcutis varies from person to person depending upon nutritional state, while the thickness of the corium and the epidermis (with stratum corneum) is above all dependent upon the location on the body. The epidermis is only between 75 and 150 μm thick. The stratum corneum which is crucial for skin absorption measures on the eye-lid less than 10 μm and on the palms of the hands and soles of the feet up to 1 mm whereby, however, the thickness of the horny layer alone is not decisive for the barrier function (8). The so-called integumentary appendages (hair follicles, excretory ducts of sweat glands, sebaceous follicles) also belong to the skin. All of these pores together only amount, however, to approx. 0.1 % of the total surface of the horny layer (20). The passage of foreign substances through these pores only plays a role for substances which otherwise penetrate the stratum corneum very slowly (19). As the barrier function of the horny layer is absent in the depth of the hair follicle it can easily be understood that results of exposure experiments with furry animals can at best be compared to the absorption behaviour of the human scalp but not as a matter of course with the absorption conditions on the human skin.

The epidermis regenerates completely under physiological conditions depending on the body locality within 45 to 75 days (22). In the basal cell layers of the epidermis where mitosis is frequent, the keratinocytes (basal cells) are formed continuously, migrate with morphological and biochemical changes into higher cell layers (prickle-cells, granular cells) and remain after the ceasing of their metabolism for 2 to 3 weeks (20) as a "dead" horny layer until they are released continuously as cutaneous scales into the environment (desquamatio insensibilis). The epidermis is therefore sometimes described as the largest holocrine gland of the body.

The appearance of the skin and its barrier function are determined decisively by the stratum corneum (horny layer). Recent models compare its structure to that of a brick wall in which the dead, protein-rich, hydrophilic keratinocytes represent the bricks and the lipid-rich hydrophobic intercellular substance represents the mortar (3). The composition and function of this intercellular substance (4) account for the fact that mathematical diffusion models which regard the barrier layer of the skin as a simple membrane have only limited applicability for describing skin absorption (8). Changes

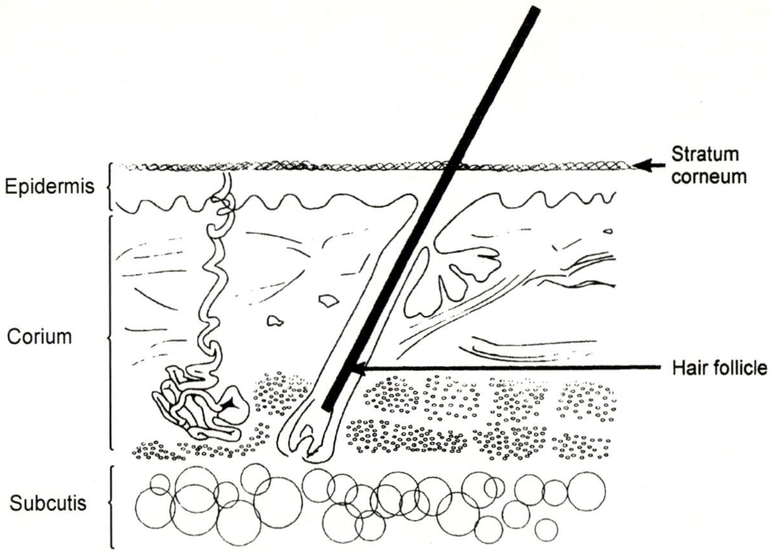

Fig. 1: Diagram of the human skin

in the structures within the stratum corneum as a result of skin disease (e. g. with florid psoriasis vulgaris, chronic eczema, ichthyosis) or the partial loss of the upper layers of the epidermis (e. g. with burns, acute eczema, bullous dermatosis) is expected to cause a significant disturbance in or complete loss of the barrier function.

Even under physiological conditions it must be assumed that skin absorption is very variable. An increased danger due to substances absorbable through the skin can be expected during work at high temperatures with the associated increased blood supply to the skin, also when there are changes in the horny layer due to permanent work in damp conditions, occlusion (e. g. rubber gloves), microtraumas of the skin and contact with certain areas of the body (e. g. the scrotal area due to contaminated underwear).

A significant modification of the dermal absorption through personal washing habits is conceivable although systematic investigations have not yet been carried out. For example before the use of the antiscabietic γ-hexachlorocyclohexane, a cleansing bath for defatting the skin is recommended so that the dermal absorption of the medicament is reduced (1, 9, 10, 18) although according to the producer no studies are available which support this hypothesis (personal communication). In contradiction with this in TRGS 150 (direct skin contact) there is the statement that when the skin is treated with defatting substances (soap, tensides, solvents) before exposure to hazardous substances absorbable through the skin, an increased danger must be assumed as defatting the skin causes increased uptake (TRGS 150).

It is also conceivable that skin care products influence dermal absorption. In the drug description leaflet for an antiscabietic (γ-hexachlorocyclohexane) there is the statement that "the simultaneous use of cosmetics … (is) to be refrained from as these can promote the unwanted uptake of the … active substance into the blood stream". This recommendation seems to be justified as some of the frequently used ointment bases (e. g. propylene glycol, isopropanol, glycerin, cetyl alcohol) are used to promote the penetration of the active substances from transdermal therapeutic systems (TTS) through the skin (21). The penetration of hydrocortisone has been shown to be promoted by the urea used in many skin medications (24). The question put to the producers of TTS as to whether body care and cleansing products could influence the absorption of active substances could not be answered on the basis of the results of systematic investigations (personal communication). In this context in particular the use of skin protection creams (barrier creams) in the factory during contact with substances absorbable through the skin is of occupational medical relevance. Systematic investigations with suitable in vitro models or of endangered collectives which take into consideration these variables affecting the percutaneous absorption of substances do not seem to be available.

These considerations of skin physiology and pharmacology show that the actual potential danger associated with skin absorption will not be ascertainable on an exclusively theoretical basis. The hazardous substances can be absorbed percutaneously both via direct skin contact and also from the atmosphere. Biological monitoring can determine the resulting total internal exposure and thus is the only valid method to estimate the individual risk.

2.3 Skin absorption after direct skin contact

Severe intoxications caused by direct contact with substances absorbable through the skin (e. g. insecticides, explosives, acrylamide, phenol) are frequent and the clinical relevance of the uptake of a hazardous substance through the skin is indisputable. Experts consider that the dermal absorption of many insecticides is of such overwhelming importance that threshold limit values should not be set for exposure by inhalation (25).

The danger of direct skin contact with hazardous substances is also a well-known problem in practical occupational medicine and has found its legal consequences in the already mentioned TRGS 150 "Direct Skin Contact with Hazardous Substances". In particular with substances with low vapour pressures (e. g. dimethylformamide, glycol ethers) the adherence to air threshold limit values is usually no problem. A significant internal exposure of employees can however result from the percutaneous uptake via the skin of both hands and forearms (ca. 2000 cm^2). Lauwerys et al. were able to show convincingly in an experiment the relevance of the dermal absorption of dimethylformamide (DMF). In the first phase of the experiment the employees worked with gloves impermeable to DMF and in the second without skin protection but with respirators. The results of biological monitoring showed that the internal exposure with respirators and no skin protection was three times as high as in phase 1 when gloves were worn. Due to gastro-intestinal problems caused by the exposure to DMF the second phase of the experiment had to be terminated prematurely (15). Therefore, when there is direct skin contact with dimethylformamide skin absorption must be regarded as the important route of absorption in contradiction to table 1.

2.4 Skin absorption from the atmosphere

The whole alveolar absorption surface has an area of approx. 60 m^2, the skin surface area, however, only 1.5 to 2 m^2. Nevertheless, skin absorption of a hazardous substance from the atmosphere can cause a significant internal exposure, as has been shown in laboratory experiments by Johanson and Borman. They exposed volunteers for two hours to 50 ppm 2-butoxyethanol (BE) entirely by inhalation via a respirator. After an appropriate pause the test persons, equipped with compressed-air respirators, then spent two hours in an exposure chamber (50 ppm BE) dressed only in shorts. The gas chromatographic determination of BE in the capillary blood of the volunteers carried out at frequent intervals showed a three to four-fold higher internal exposure to BE during and after purely dermal exposure than during inhalation. It was estimated that when direct skin contact was excluded and persons exposed freely via the atmosphere, dermal absorption accounted for 75 % of the total substance absorbed; the usefulness of protection with only a respirator was called into question (11).

Dermal absorption from the atmosphere must therefore be taken into account with all substances absorbable through the skin if breathing protection is worn but no protective clothing. Also in this case the actual exposure of the employee can only be determined with biological monitoring.

It must also be assumed that dermal absorption of hazardous substances from the atmosphere persists after the end of exposure when the substances are no longer inhaled. Both in the clothing of the employee and also on the surface of his skin and in the stratum corneum, whose depot function is well-known from pharmacological investigations (13), the hazardous substance can be found in concentrations which depend upon the substance involved and cause a continuing uptake. A change of clothes and cleansing of the skin e. g. by showering can decrease the individual exposure by removing particles from the surface of the skin (14).

2.5 Skin absorption of carcinogenic substances for which no TRK or EKA can be formulated

As explained above, for carcinogenic substances no threshold limit values considered safe for health can be evaluated. The advantages of biological monitoring for individual protection should particularly benefit those persons who deal with particularly dangerous substances, i.e. with carcinogenic substances. This is possible for carcinogenic substances which have both TRK values and EKA. Difficulties occur, however, if there are no appropriate TRK values or when due to low vapour pressures only extremely low concentrations of the hazardous substance are to be expected in the air and thus EKA cannot be established. This is the situation, for example, with many aromatic amines which on the one hand are characterised by low vapour pressures but on the other hand are absorbed particularly well through the skin (e. g. benzidine, 3,3-dichlorobenzidine, 2-naphthylamine). There is a similar problem with substances (e. g. 4,4'-diaminodiphenylmethane) for which data are available for internal exposure only at air values greatly below the TRK value. Often extrapolation to higher values is not possible because with increasing air concentration the additional skin absorption leads to exponential and no longer calculable increases in the internal exposure (Lewalter, personal communication). Naturally in these cases no EKA values can be established. Nevertheless biological monitoring should not be neglected. Only biological monitoring can determine exposures to hazardous substances via routes of absorption other than the bronchial tract. On the basis of the information obtained, recommendations can then be worked out to make it possible to evaluate exposures under the specific exposure conditions.

The considerations presented here show that the practical and well-validated instrument of biological monitoring in occupational medical health care should be implemented in particular under conditions where various routes of absorption and/or an individual variability of the internal exposure are expected. These aspects have now been taken into account in considerations on the general position of biological monitoring in the form of a technical guideline for hazardous substances (16).

3 References

1. Braun-Falco, O., Plewig, G., Wolf, H. H.: Dermatologie und Venerologie. Springer Verlag, Berlin-Heidelberg 1984
2. Deutsche Forschungsgemeinschaft: MAK- und BAT-Werte-Liste 1992. Maximale Arbeitsplatzkonzentrationen und Biologische Arbeitsstofftoleranzwerte. VCH Verlagsgesellschaft, Weinheim 1992
3. Fartasch, M., Hühner, A., Tepe, A., Funk, U., Diepgen, T. L.: Hautphysiologische Untersuchungsmethoden in der Berufsdermatologie. Allergologie *16* (1993 a), 25–34
4. Fartasch, M., Bassukas, I. D., Diepgen, T. L.: Structural relationship between epidermal lipid lamella, lamellar bodies and desmosomes in human epidermis: an ultrastructural study. Br. J. Dermatol. *128* (1993 b), 1–9
5. Fiserova-Bergerova, V., Pierce, J. T., Droz, P. O.: Dermal absorption potential of industrial chemicals: criteria for skin notation. Am. J. Ind. Med. *17* (1990), 617–635
6. Forth, W., Henschler, D., Rummel, W.: Pharmakologie und Toxikologie. Bibliographisches Institut AG, Zürich 1983
7. Grandjean, P., Berlin, A., Penning, G. M.: Preventing percutaneous absorption of 1.1.1-trichloroethane (2). Am. J. Ind. Med. *14* (1988), 97–107
8. Grandjean, P.: Skin Penetration. Taylor & Francis, London 1990
9. Hermann, W. P., Steigleder, G. K.: Scabies. Dtsch. Med. Wschr. *92* (1967), 1557–1561
10. Hornstein, O. P.: Bakterielle und parasitäre Dermatosen. In: Hornstein, O. P., Nürnberg, E. (eds.): Externe Therapie von Hauterkrankungen. Thieme Verlag, Stuttgart 1985
11. Johanson, G., Borman, A.: Percutaneous absorption of 2-butoxyethanol vapour in human subjects. Br. J. Ind. Med. *48* (1991), 788–792
12. Kennedy, G. L., Brock jr., J. W., Banerjee, A. K.: Assignment of skin notation for threshold limit values chemicals based on acute dermal toxicity. Appl. Occup. Environ. Hyg. *8* (1993), 26–30
13. Koch, E.: Pharmakokinetische Aspekte von Dermatika, In: Hornstein, O. P., Nürnberg, E. (eds.): Externe Therapie von Hauterkrankungen. Thieme Verlag, Stuttgart 1985
14. Koizumi, A.: Experimental evidence for the possible exposure of workers to hexachlorobenzene by skin contamination. Br. J. Ind. Med. *48* (1991), 622–628
15. Lauwerys, R. R., Kivits, A., Lhoir, M., Rigolet, P., Houbat, D., Buchet, J. P., Reets, H. A.: Biological surveillance of workers exposed to dimethylformamide and the influence of skin protection on its percutaneous absorption. Int. Arch. Occup. Environ. Health *45* (1980), 189–203
16. Miksche, L. W., Meyer, G., Perlebach, E.: Biomonitoring im Rahmen des Gesundheitsschutzes am Arbeitsplatz. Arbeitsmed. Sozialmed. Umweltmed. *28* (1993), 388–391

17. Scansetti, G., Pilatto, G., Rubino, G. F.: Skin notation in the context of workplace exposure standards. Am. J. Ind. Med. *14* (1988), 725–732
18. Schmeel, A., Lutz, G., Bauer, R.: Die Krätze. Therapiewoche *34* (1984), 5699–5706
19. Scheuplein, R. J., Bronaugh, I. H.: Percutaneous absorption. In: Goldsmith, L. A. (ed.): Biochemistry and Physiology of the skin. Oxford University Press, Oxford 1983
20. Stüttgen, G.: Die biologische Funktion des Stratum corneum. Pharmakokinetische Bedingungen für die topische Applikation. In: Rietbrock, N. (ed.): Die Haut als Transportorgan für Arzneistoffe. Steinkopff Verlag, Darmstadt 1990
21. Szycher, M.: Recent advances in transdermal delivery Systems. J. Biomat. Appl. *1* (1986), 207–238
22. Tronnier, H., Kresken, J., Jablonski, K., Komp, B.: Haut und Beruf. Grosse Verlag, Berlin 1989
23. Wiechers, J. W.: The barrier function of the skin in relation to the percutaneous absorption of drugs. Pharm. Weekbl. (Sci) *11* (1989), 185–198
24. Wohlrab, W., Taube, K. M., Kuchenbecker, I.: Penetration und Wirksamkeit von Hydrokortison bei reduzierter Konzentration im Vehikel. Zeitschrift für Hautkrankheiten *65* (1990), 534–537
25. World Health Organization (WHO): Recommended health-bases limits in occupational exposure to pesticides. Report of a WHO study group. Technical Report Series *677*, Genf 1982

Authors: H. Drexler, J. Angerer, K.-H. Schaller, G. Lehnert

BAT Value Documentation

Acetylcholinesterase Inhibitors

BAT

Inhibition of the erythrocytic acetylcholinesterase activity:

	Reference value	BAT
Time of activity determination	Before exposure	After exposure
Value	Individually variable ≙ 100 %	Reduction of the activity to 70 % of the reference value

Sampling: end of shift

Date of evaluation 1985

Substances

Organic phosphoric acid esters (phosphates)
Carbamic acid esters (carbamates)
Organic sulphuric acid esters (sulphates)
Sulphonic acid esters (sulphonates)

Synonyms

Alkyl phosphates
Alkyl sulphates
Carbamates

Formula

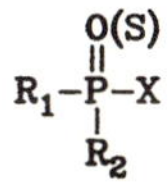

R_1 = alkoxy
R_2 = alkoxy, alkyl, dialkylamido
X = halogen, cyanide, phenoxy, disubstituted pyrophosphate

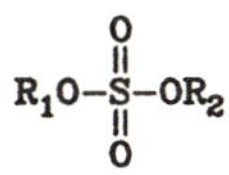

R_1, R_2 = alkyl

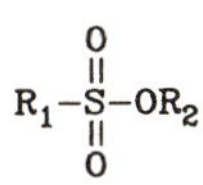

R_1 = alkyl, aryl
R_2 = alkyl

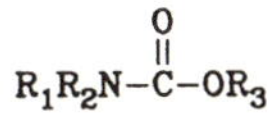

R_1 = H, alkyl,
R_2, R_3 = alkyl

Organic phosphates, organic sulphates, sulphonates and carbamates are characterised by a common spectrum of effects. The inhibition of erythrocyte acetylcholinesterase has proved suitable as an indicator of the toxicity of these substances.

1 Biochemistry of Cholinesterases

The cholinesterases are divided into two subgroups (9, 31): *1.* Acetylcholinesterase in the narrower sense (AChE) which is present bound to cell structures in the regions of the cholinergic synapses, the grey substance of the central nervous system, the autonomic ganglions, the pre and postganglionic sympathetic synapses, the motor end-plates in muscles, as well as the parasympathetic postganglionic synapses and in erythrocytes. As part of the acetylcholine system this enzyme has the physiological function of splitting the transmitter acetylcholine (ACh) rapidly into choline and acetic acid and thus to inactivate it (29). *2.* Cholinesterases (ChE), also called pseudocholinesterases, form a group of isoenzymes. They are less specific and present ubiquitously in the organism. Their concentration in the plasma is 7–9 mg/l. The pseudocholinesterases in the plasma differ with regard to their substrate specificities, pH optima, electrophoretic mobilities and kinetics (22). Their physiological function is unknown. A pharmacological function is the splitting of certain drugs such as procaine,

Tab. 1: Nomenclature and characteristic properties of acetylcholinesterase and pseudocholinesterase (according to (31))

Recommended name	Acetlycholinesterase	Cholinesterase
Abbreviation	AChE	ChE
Systematic name	Acetylcholine-acetyl-hydrolase	Acylcholine-acyl-hydrolase
EC code number	3.1.1.7	3.1.1.8
Synonyms	Specific cholinesterase, genuine cholinesterase	Unspecific cholinesterase, pseudocholinesterase, plasma or serum cholinesterase, butyrylcholinesterase, benzoylcholinesterase
Optimum substrates	Acetylcholine	Butyrylcholine, benzoylcholine
Effect of high acetylcholine concentrations (excess substrate)	Inhibition	No inhibition
Benzoylcholine	Not a substrate	Optimal substrate
Acetyl-β-methylcholine	Substrate	Not a substrate
Isoflurophate	Weak inhibition	Strong inhibition

succinylcholine or succinylbischoline (28) and acetylsalicylic acid as well as the detoxification of phosphates and carbamates. In table 1 the characteristic differences between acetylcholinesterase and pseudocholinesterase have been listed. Unlike pseudocholinesterase, acetylcholinesterase is an essential enzyme with a high substrate specificity (37).

1.1 Enzymatic Reaction Mechanism of Acetylcholinesterase and Pseudocholinesterase

Currently available results indicate that there are no fundamental differences between the catalytic centres of the two cholinesterases (21, 40).

They both contain at least two enzymatically effective regions (45) which are described as the "anionic" and "esteratic" centre (Fig. 1). In the negatively charged anionic centre the enzyme-substrate interaction begins and forms a labile enzyme-substrate complex (21) (Fig. 1). In the esteratic centre the reaction partners form a covalent bond with the carbonyl carbon and a dipole bond with the carbonyl oxygen of the acetylcholine. After intramolecular electron transfer and rearrangement the choline part dissociates from the anionic centre forming as an intermediate the acetylated enzyme (acetylation step). The acetylcholinesterase is once again available unchanged as an enzyme ready for reaction (21). The rate-limiting step is the deacetylation as it takes place more slowly than the dissociation of the choline. The half-time for the acetylation is about 7 µs, the half-life of the acetylated esterase, however, 42 µs (42).

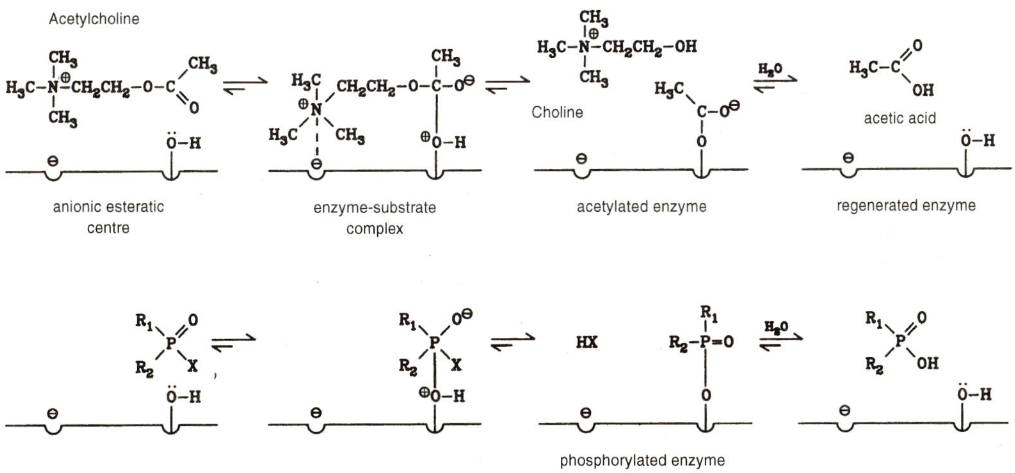

Fig. 1: Diagram of the course of reaction between acetylcholine or a phosphate and AChE. Short arrows mean slow, extremely slow or insignificant reaction speeds (modified from (29))

1.2 Inhibition of Acetylcholinesterase and Pseudocholinesterase

From the clinical standpoint, the inhibition of both cholinesterases is classified as "reversible" or "irreversible" according to the extent and rate of the dissociation of the inhibitor from the enzyme, that is, of the reactivation of the enzyme (37). The acetylcholinesterase is inhibited by a great variety of substances with chemically very different structures (17). In addition to drugs whose desired effect is based on reversible inhibition of the acetylcholinesterase (e.g. neostigmine) (24) there are those whose (reversible) AChE inhibition is only a side effect of their actual pharmacological effect (pethidine, immunosuppressants, cytostatic agents, etc. (36)). The classic examples of the so-called irreversible inhibitors are the organic phosphates (alkylphosphates) of which many more than 2000 have been synthesised to date (26). Many of these lipophilic substances are taken into the organism readily via the skin as well as by inhalation. They form an enzyme-inhibitor complex by phosphorylation of the enzyme.

The inhibitors of the cholinesterases are grouped according to their mechanism of action (39):

1. Prosthetic inhibitors which react with the anionic enzyme centres.

The reversible competitive inhibitors are assigned to this group (e.g. neostigmine).

2. Oxydiaphoric inhibitors which can have either reversible or irreversible effects (17a):

a. Carbamates: drugs, e.g. physostigmine, pyridostigmine and insecticides, e.g. carbaryl and isolan® (14). The symptoms of carbamate intoxication are basically similar to those with alkylphosphate intoxication. However, they subside more quickly. Death is seldom (18).

b. Organic phosphates (alkylphosphates) (41): There are at present 50 different substances in practical use. They are esters or amides of phosphoric acid, thiophosphoric acid, phosphonic acid or phosphinic acid. For example: parathion (insecticide), nitrostigmine (miotic), bromophos (insecticide), demeton (systemic insecticide, acaricide), dichlorvos (insecticide) and dimethoate (insecticide) (14). The different toxicity of the phosphates is mainly a result of the fact that some esters are metabolised and thus detoxified more quickly by warm-blooded animals but more slowly by insects. Furthermore, the differences between species are important in the dimensions of the catalytic centre of AChE (10).

Some alkylphosphates can only react with the cholinesterases after metabolism (indirect inhibition). The toxicity of all thiophosphates is increased by oxidation of the P = S group to form a P = O group. For example, during parathion exposure (E 605) paraoxon forms the irreversible enzyme-substrate complex, but not parathion.

The efficacy of phosphates can be varied by changes in the side chain. Thus malathion is quickly broken down with elimination of alcohol to form the acid with a polar side chain which is no longer capable of inhibiting AChE. As insects do not have the appropriate esterases, malathion is a sufficiently effective insecticide despite its low toxicity for warm-blooded animals. If these esterases of warm-blooded animals are inhibited by other alkylphosphates the unchanged malathion is also highly toxic in warm-blooded animals (superadditive combination effect). Often activation and inactivation processes take place simultaneously in the same substrate so that the effective toxicity is a result of the individual rates of the two metabolic processes (18).

c. Organic sulphuric acid esters (alkyl sulphates) and sulphonates (20): There is, so far, only little practical experience with these compounds.

All acid-transferring inhibitors react with the esteratic centre of the acetylcholinesterase in a manner analogous to the reaction of the physiological substrate acetylcholine and transfer the acid ester (carbamic acid, phosphoric acid, sulphuric acid) onto the enzyme to form a very hydrolysis-resistant ester bond with the serine hydroxy group (1).

The stability of the enzyme-ester bond of the alkylphosphates should be directly proportional to the size of the alkyl substituents on the phosphorus atom (36). With organic sulphuric acid esters the formation of an initially reversible enzyme-inhibitor complex has so far not been observed (20). Unlike the reaction of the acetylcholinesterase with acetylcholine, its reaction with an inhibitor leads to a carbamoylated, phosphorylated or sulphated enzyme which is much more stable than the acetylated enzyme.

During hydrolytic cleavage of acetylcholine the acetylated enzyme reacts within fractions of a second with water (43). However, the hydrolysis of the acetylated enzyme takes several minutes after inhibition with carbamates (15), with alkylphosphates up to several weeks (40) and with methylsulphates even several months (2).

After reaction with reversible inhibitors (carbamates) the initial activity returns within a few hours, with irreversible blockers (alkylphosphates) the esteratic activity is mainly recovered after new synthesis (within a period of several weeks) (27).

2 Critical Toxicity

The clinical picture after intoxication with AChE inhibitors is produced exclusively by the inhibition of the bound acetylcholinesterase at the cholinergic synapses. The acute inhibition of the acetylcholinesterase leads to an accumulation of the transmitter substance in the tissues ("endogenous acetylcholine intoxication"). Depending on the severity of the inhibition, the sequelae vary from the characteristic symptoms of parasympathetic stimulation to a vegetative, motor and central nervous cholinergic crisis (Tab. 2) (19, 27a).

Tab. 2: Symptoms of an ACh intoxication due to alkylphosphates ("endogenous ACh intoxication") (from (31)).

Muscarine-like effects (Parasympathetic nervous system)	Nicotine-like effects (motor end-plates, autonomous ganglions, adrenal medulla)	Central nervous effects
Miosis Accommodation disorders Increased secretion from the tear, sweat and salivary glands, as well as the mucous glands in the tracheobronchial tract Stimulation of the smooth muscles of the bronchial tract (dyspnoe, bronchiospasm), of the gastro-intestinal tract (vomiting, diarrhoea, colic) and of the urogenital system (urinary incontinence) Vasodilation Bradycardia	Initially or during mild intoxication: stimulation Later during severe intoxication: inhibition Muscle twitching in the eye lids and tongue Occasional hypertension and hyperglycaemia Hyperglycaemia Generalised fasciculation Trembling, convulsions Muscular weakness, respiratory paralysis	Agitation, confusion Nightmares Headache Stupor, coma Convulsions EEG: Initially desynchronisation, then increase in the frequency and amplitude of the discharges, potential for convulsions Respiratory and circulatory depression

The first clinically significant symptoms caused by inhibition of the acetylcholinesterase occur only when its activity has decreased markedly under 50 % (8). Critical conditions of intoxication are observed after decrease of the acetylcholinesterase activity to 20 % of the individual norm value (5). After intoxication with organic phosphates persistent damage to sensory and motor nerves has been described occasionally. After intake of large amounts of alkylphosphates death occurs rapidly due to central respiratory paralysis (35, 44).

Chronic exposure to AChE inhibitors occurs in industry and agriculture. The danger from chronic intake of inhibitors is relatively small as many of these substances are rapidly hydrolysed in the body and thus lose their efficacy. There may, however, be additive or potentiated effects of mixed exposure to different cholinesterase inhibitors (4). Impairment of parenchymatous organs after chronic exposure to cholinesterase inhibitors has not been observed (25). On the other hand, depending upon the degree of accumulation, muscarinic and central nervous effects have been observed (44).

3 Exposure and Effects

The relationships between external exposure (concentration of the AChE inhibitor at the workplace), internal exposure (concentration of the AChE inhibitor or the metabolite causing AChE inhibition in body fluids, organs etc.) and effect (cholinesterase inhibition) are not only dependent upon the kind of cholinesterase inhibitor (see section 1) but to a great extent also on the individual (genetic) disposition of the person exposed (13).

As the possibility of absorption via the skin plays an important role with many organic phosphates a correlation between the ambient air concentration at the workplace and biological parameters is only very rarely possible in a specific case. Furthermore, in the evaluation of intoxication symptoms, particularly in the area of agricultural usage, it must be taken into account that the commercial preparations contain additives such as solvents and emulsifiers (e.g. dimethylformamide, chlorobenzene, *o*-dichlorobenzene, glycol ether) whose toxicity can be clinically more important.

The symptoms caused by cholinesterase inhibitors and their quantitative evaluation are discussed in section 2 “Critical Toxicity”.

4 Selection of the Indicators

Determination of acetylcholinesterase inhibition provides a direct measure of a biological parameter for the damage, independent of the chemistry of the acetylcholinesterase inhibitor. In particular the determination of the inhibition of erythrocyte acetylcholinesterase has proved suitable for quantitative purposes for biological monitoring (23).

In practice use of the serum cholinesterase activity as evidence of exposure to alkylphosphates is limited by the considerable range of normal serum pseudocholinesterase activity (6, 12). Thus, even within unexposed collectives the highest values measured are nearly twice as high as the lowest (16). The individual acetylcholinesterase or pseudocholinesterase activities seem, however, to be very constant over many years (38). This is due in part to the fact that in addition to persons whose genetically determined pseudocholinesterase activity is normal (homozygotes), there are both persons with a slightly decreased pseudocholinesterase activity compared to the norm (heterozygotes) and also homozygous persons with extremely low activity (13). Cases have been described with hereditary pseudocholinesterase activity whose value is zero ("silent gene") (33).

During many illnesses, in particular when the liver parenchyma is damaged, the pseudocholinesterase activity is below the norm (11). During other illnesses, e.g. kidney disease, it can be considerably increased (34). Furthermore, all cholinesterases can be inhibited not only by organic phosphates, organic sulphates, sulphonates and carbamates but also to a greater or lesser extent by drugs (physostigmine, prostigmine) (12).

In particular pseudocholinesterases are subject to a very steep daily gradient so that even with unexposed persons the individual morning values are generally above the corresponding evening values (6). Insufficient protein synthesis (decreased new synthesis of the enzyme) in the liver parenchyma cells can decrease the cholinesterases in the plasma or serum by 30 to 40 % (30). In order to be able to differentiate these effects in particular from possible chronic exposure to cholinesterase inhibitors, the reliable determination or evaluation of the erythrocytic acetylcholinesterase activity is recommended since this is subject to much smaller daily fluctuation.

5 Methodology

For the determination of acetylcholinesterase activity both venous and capillary blood samples are suitable. For the measurement of the acetylcholinesterase activity washed erythrocytes, or if necessary whole blood is used. The quantitative determination takes place with a photometric test. The loose-leaf collection "Analysen in biologischem Material" ("Analyses of Hazardous Substances in Biological Materials") (3) contains tested and practicable methods.

6 Evaluation of the BAT value

According to the data available at present, only the inhibition of acetylcholinesterase leads to symptoms of internal acetylcholine accumulation with the clinical findings of muscarine-like, nicotine-like and central nervous effects (35). The WHO tolerates contact with acetylcholinesterase inhibitors if their cholinesterase inhibition in plasma is limited to 50 % of the individual cholinesterase activity before contact with the inhibitor (7). As detailed in sections 2 and 4, however, only the acetylcholinesterase activity, which can be determined in the isolated erythrocytes, can be used as a valid indicator of effect. If the erythrocyte acetylcholinesterase activity decreases after exposure to the inhibitor to below 30 % of the individual reference value, clinical manifestations of an internal acetylcholine intoxication are to be expected (7, 8, 14, 23, 25, 35, 44, 46). As a BAT value a

reduction by 30 % of the acetylcholinesterase activity in the isolated erythrocytes to 70 % of the individual reference value

has been set.

It limits equally all effects due to phosphates, sulphates, sulphonates, carbamates and their mixtures. Such a threshold limit value is in accordance with the recommendations of the WHO (7).

7 Interpretation of the Data

The above documentation of the BAT value presupposes for its correct usage knowledge at least of the individual reference value for the acetylcholinesterase activity. This should be determined at least once a year after an exposure-free interval.

In order to determine reliably representative and individually comparable acetylcholinesterase activities the blood sampling for the determination of the AChE activity should preferably be carried out after 3 consecutive working days. The blood sampling should be carried out under contamination-free conditions, preferably after washing and in private clothing, in the medical department. The isolation of the erythrocytes (centrifugation and washing) should take place no later than 3 hours after blood sampling in order to avoid haemolysis.

It should be borne in mind in the evaluation of the values obtained that women can have lower acetylcholinesterase activity values than men (32) and that, furthermore, some medicines can change the activity of the acetylcholinesterase. Thus if the AChE activity values are conspicuous, a medical anamnesis should be carried out (see section 1.2).

8 References

1. Aldridge, W. N.: Organophosphorus compounds and carbamates, and their reaction with esterases. Br. Med. Bull. *25* (1969), 236
2. Alexander, J., Wilson, J. B., Kitz, R. J.: The reactivation of acetylcholinesterase after inhibition by methansulfonic esters. J. Biol. Chem. *238* (1963), 741
3. Henschler, D. (ed.): Analytische Methoden zur Prüfung gesundheitsschädlicher Arbeitsstoffe. Band 2: Analysen in biologischem Material. VCH Verlagsgesellschaft, Weinheim 1976
4. Barnes, J. M.: Toxicité pour l'homme de certains pesticides. Organis. mond. Santé, Geneve 1954
5. Barnes, J. M., Davies, D. R.: Blood Cholinesterase Levels in Workers exposed to Organo-Phosphorous Insecticides. Brit. med. J. (1951/II), 816

6. Brown, S. S., Kalow, W., Pilz, Whittaker, M., Woronick, C. L.: The Plasma Cholinesterases: A new Perspective. Advances in Clinical Chemistry *22* (1981), 1–123
7. World Health Organization (WHO): Data Sheet on Pesticides No. 6: Parathion. WHO, Geneva 1975
8. Dubois, K. P.: Potentiation of the Toxicity of Insecticidal Organic Phosphates. Arch. Ind. Health *18* (1958), 488
9. Enzyme Nomenclature. Recommendations (1964) of the International Union of Biochemistry on the Nomenclature and Classification of Enzymes. Elsevier, Amsterdam 1965
10. Fest, C., Schmidt, K. J.: Insektizide Phosphorsäureester. In: Buechel, K. H.: Pflanzenschutz und Schädlingsbekämpfung. Thieme, Stuttgart 1977, pp. 22–60
11. Fintelmann, V., Lindner, H.: Diagnostische Bedeutung der Serumcholinesterase bei Lebererkrankungen. Dtsch. med. Wschr. *95* (1970), 469
12. Friedberg, K. D., Sakai, F.: Spezifischer Nachweis von Vergiftungen mit Alkylphosphaten (E 600, E 605, Systox) in Blut und Hirngeweben mit Hilfe eines ferment-reaktivierenden Antidots (Pyridin-Aldoxim-Methyl-jodid, PAM). Dtsch. Z. Ges. Gerichtl. Med. *47* (1958), 580
13. Goedde, H. W., Doenicke, A., Altland, K.: Pseudocholinesterasen. Springer, Berlin 1967
14. Henschler, D. in: Forth, W., Henschler, D., Rummel, W. (eds.): Allgemeine und spezielle Pharmakologie und Toxikologie. 4th ed. Bibl. Inst., Mannheim 1983, pp. 674–679
15. Hobbinger, F.: Inhibition of cholinesterases by irreversible inhibitors in vitro and in vivo. Br. J. Pharmacol. *6* (1951), 21
16. Kalow, W.: Pharmacogenetics. Saunders, Philadelphia 1962
17. Karczmar, A. G.: Pharmacologic, toxicologic and therapeutic properties of anticholinesterase agents. In: Roots, W. S., Hoffmann, F. G. (eds.): Physiological Pharmacology, Vol. III. Academic Press, New York 1967

17a. Karczmar, A. G., Usdin, E., Wills, J. H.: Anticholinesterase Agents. Vol. 1. Pergamon Press, Oxford 1970

18. Kimmerle, G., Lorke, D.: Die Toxikologie der insektiziden Phosphorsäureester. Pflanzenschutznachrichten Bayer *21* (1968), 111–142
19. Kitz, R. J.: The chemistry of anticholinesterase activity. Acta Anaesthesiol. Scand. *8* (1964), 197
20. Kitz, R. J., Wilson, J. B.: Esters of urethansulfonic acid as irreversible inhibitors of ACHE. J. Biol. Chem. *237* (1962), 2345
21. Korolkovas, A.: Grundlagen der molekularen Pharmakologie. Thieme, Stuttgart 1974
22. La Motta, U. R. V., Woronick, C. L.: Molecular heterogenety of human serum cholinesterase. Clin. Chem *17* (1971), 135
23. Lauwerys, R. L.: Industrial Chemical Exposure. Guidelines for Biological Monitoring. Davis: Biomedical Publ. (1983), 118–121
24. Mayrhofer, O.: Erfahrungen mit Galanthamin (Nivalin) als Antagonist der Relaxantien vom Curaretyp. Bull. Schweiz. Akad. Med. Wiss. *23* (1967), 48
25. Moeschlin, S.: Cholinesterase-Blocker. In: Klinik und Therapie der Vergiftungen. Thieme, Stuttgart 1972, pp. 354–362
26. Motulsky, A. G.: Pharmacogenetics. Prog. Med. Gent. *3* (1964), 49
27. Namba, T.: Cholinesterase inhibition by organophosphorous compounds and its clinical effects. Bull. WHO. *44* (1971), 289–307

27a. O'Brien, R. D.: Insecticides. Action and Metabolism. Academic Press, New York 1967

28. Pilz, W., Boo, A. T., Stelzer, E.: Quantitative Verfolgung des Wirkungsverlaufes von Bernsteinsäuredicholinester und Bernsteinsäuremonocholinester auf esterspaltende Enzyme des menschlichen Blutes. Z. klin. Chem. *5* (1967), 134
29. Potter, L. T.: Acetylcholine, choline, acetyltransferase, acethylcholinesterase. In: Lajtha, A. (ed.): Handbook of Neurochemistry, Vol. VI. Plenum Press, New York 1970
30. Richterich, R., Colombo, J. P.: Klinische Chemie, Theorie, Praxis, Interpretation. 5. Karger, Basel, München, New York, 4. Aufl. (1978), 392–510 particularly 498 505
31. Schuh, F. T.: Über Cholinesterasen – eine anaesthesiologisch wichtige Enzymgruppe. Anaesthesist *25* (1976), 501
32. Sidell, F. R., Kaminskis, A.: Influence of age, sex and oral contraceptives on human blood cholinesterase activity. Clin. Chem. *21* (1975), 1393
33. Simpson, N. E., Kalow, W.: The "silent" gene for serum cholinesterase. Amer. J. hum. Genet. *16* (1964), 180

34. Stefenelli, N.: Die Korrelation zwischen dem Albumingehalt und der Cholinesteraseaktivität des Blutserums und ihre Beurteilung bei Albuminverlust. Klin. Wschr. *19* (1961), 1019
35. Valentin, H., Lehnert, G., Petry, H., Weber, G., Wittgens, H., Woitowitz, H. J.: Arbeitsmedizin. Vol. 2.: Berufskrankheiten, BK-Nr.: 1307. Thieme, Stuttgart 1979, pp. 90–93
36. Wang, R. J. H., Ross, C. A.: Prolonged apnoea in cancer patients receiving AB-132. Anestesiology *24* (1965), 363
37. Wassermann, O.: Acetylcholinesterase: Funktion, Hemmung und Reaktivierung. Pharmacia *23* (1968), 49
38. Wetstone, H. H., La Motta, R. V.: The clinical stability of serum cholinesterase activity. Clin. Chem. *II* (1965), 653
39. Wilson, J. B.: Acetylcholinesterase XI. Reversibility of tetraethylpyrophosphate inhibition. J. Biol. Chem. *190* (1951), 111
40. Wilson, J. B.: The active surface of the serum esterase. J. Biol. Chem. *208* (1954), 123
41. Wilson, J. B.: Inhibition of ACHE In: CIBA Foundation Symposium on Enzymes and Drug Action. Churchill, London 1962
42. Wilson, J. B., Cabib, E.: Acetylcholinesterase: enthalpies and entropies of activation. J. Am. Chem. Soc. *78* (1956), 202
43. Wilson, J. B., Harrison, M. A.: Turnover number of acethylcholinesterase. J. Biol. Chem. *236* (1961), 2292
44. Wirth, W.: Zum Nachweis der E 605-Intoxikation. Untersuchung der Cholinesterase-Aktivität im Hirn. Arch. Toxicol. *12* (1956), 125–128
45. Zeller, E. A., Bissegger, A.: Über die Cholinesterase des Gehirns und der Erythrocyten. III. Mitteilung über die Beeinflussung der Fermentreaktion durch Chemotherapeutika und durch Pharmaka. Helv. Chim, Acta *26* (1943), 1619
46. Zielhuis, R. L.: Biological monitoring. Scand. J. Work Environ. Health. *4* (1978), 1–18

Author: J. Lewalter
Approved by the Working Group: 10th May 1985

Aniline

BAT

free aniline	1 mg/l urine
aniline released from aniline-haemoglobin conjugate	100 µg/l blood

Date of evaluation 1985

Addendum 1993

Synonyms: aminobenzene, phenylamine

Formula: C_6H_7N

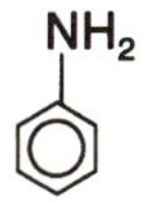

Molecular weight: 93.13

Melting point: −6.3 °C

Boiling point: 184.13 °C

Vapour pressure at 20 °C: 64 Pa (0.48 torr)

MAK: 2 ml/m^3 (ppm), 8 mg/m^3

1 Metabolism and Kinetics

Toxic effects of aniline are observed after inhalation and after oral or cutaneous exposure (6). It is quickly absorbed and its metabolites are excreted within 24 to 48 hours, predominantly via the kidneys. An accumulation of metabolites bound to haemoglobin is observed and, as a consequence of erythrocyte degradation, in the spleen (18, 22). Recent investigations on various species of animal have shown that the main route of metabolism takes place via acetylation to acetanilide and further to *N*-acetyl-

p-aminophenol (13, 27). In addition O-conjugates of *o*-aminophenol and *p*-aminophenol are found in the urine (14). Phenylhydroxylamine and nitrosobenzene are only excreted in small amounts. The main metabolic pathways of aniline are shown in figure 1. Furthermore, additional metabolic pathways of aromatic amines have been discussed recently, e. g. co-oxidation by prostaglandin synthetase (21) and conjugation reactions of hydroxamic acids (12). The relationship of *o*-hydroxylation to *p*-hydroxylation depends to a great extent upon the kind of animal and in humans, moreover, is subject to great individual variation (24). *m*-Aminophenol is only formed in small amounts. With increasing exposure to aniline the formation of aminophenol increases significantly (25).

In addition to acetylation N-oxidation takes place to a small extent. This leads to the formation of methaemoglobin mainly because phenylhydroxylamine is formed from aniline in the liver and then oxidised in the erythrocytes to nitrosobenzene (7). In the presence of oxygen in coupled oxidation reactions the bivalent iron in the haemoglobin is simultaneously oxidised to the trivalent form.

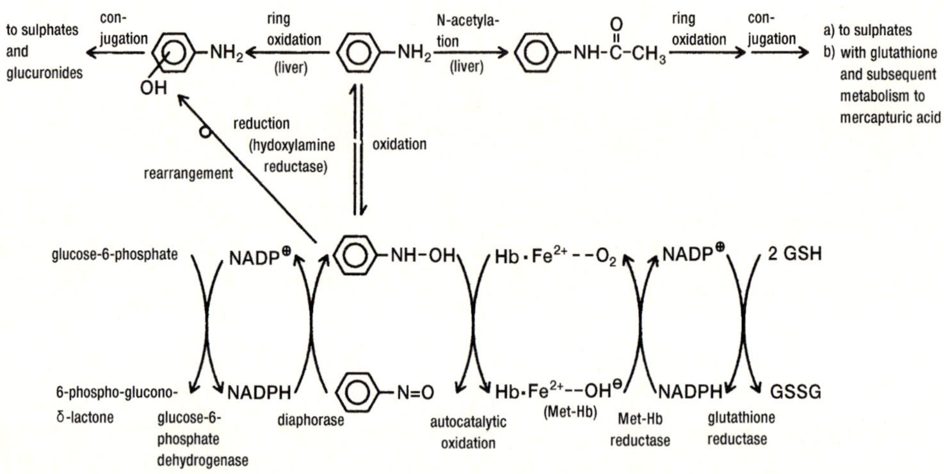

Fig. 1: Metabolism of aniline (4, 5, 7, 8)

This cycle can repeat itself several times as nitrosobenzene can be reduced to phenylhydroxylamine again by NADPH-dependent diphorases. The transformation of phenylhydroxylamine to *o*-aminophenol and *p*-aminophenol and the conjugation of the intermediates involved are considered to be detoxifying mechanisms (10). The intermediate, nitrosobenzene, can also bind covalently to SH-groups of haemoglobin and glutathione (1, 23, 28, 31) forming a sulphinic acid amide. These reactions are shown in figure 2.

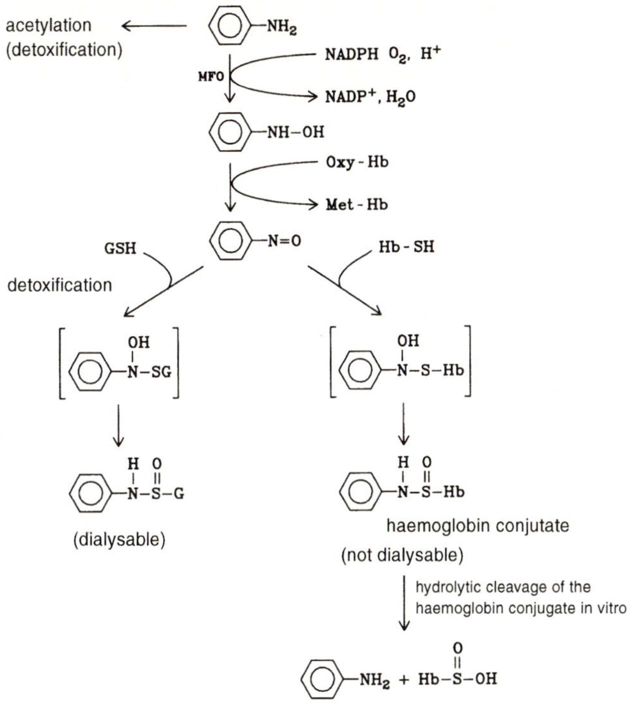

Fig. 2: Formation of aniline conjugates with glutathione and haemoglobin (according to (28)); MFO = mixed function oxygenases

2 Critical Toxicity

High concentrations of aniline in the workplace air quickly lead, i. e. within 10–30 minutes, first to central nervous effects, then via dizziness and clouding of consciousness to unconsciousness. The formation of methaemoglobin is recognisable by the blue colouring of the lips, nose, ears and nail bed. These acute toxic effects can be prevented by observance of the MAK value.

Sub-acute aniline exposures are also supposed to have effects on liver metabolism (25). The type and extent are, however, not described in any more detail. With large-scale destruction of the erythrocytes kidney damage can occur. In isolated cases changes in the white blood count and damage to the bone marrow have been described (29).

Chronic aniline exposures can lead after only slight formation of methaemoglobin and correspondingly only slight cyanosis to the formation of Heinz (inclusion) bodies and anaemia. With slow progression often only subjective complaints occur until finally occasionally hypochromic anaemia and isolated inclusion bodies can be found, usually with only slightly increased methaemoglobin levels. Manifest clinical symptoms do not occur until the methaemoglobin level is more than 10 %. Known effects in humans are described in detail in the MAK documentation for aniline (3).

2.1 Methaemoglobin Formation

For details of the pathophysiology of methaemoglobinaemia reference can be made to the current MAK documentation for aniline (3). Notes and information on the practical use of methaemoglobin determination in biological monitoring can be found in a publication by the Working Group on the problem of establishing BAT values for aromatic amines (6).

2.2 Other Effects of Aniline

In addition to the formation of methaemoglobin, exposure to aniline also causes other effects. The formation of Heinz bodies has been observed, inclusion bodies resulting from denaturation of haemoglobin. Chronic uptake of aniline can cause haemolysis and lead to anaemia. In addition to these effects on erythrocytes, neurotoxic effects can be expected.

2.3 Possible Carcinogenic Effects

For a discussion of potential carcinogenic effects of aniline and aromatic amines reference can be made to a review compiled by the Commission (6). On the basis of experiments in which aniline was fed to rats in very high doses for a prolonged period this substance is suspected of having carcinogenic potential. Therefore it is listed in category III B of the List of MAK and BAT Values (19a). The ability of aniline metabolites to react with macromolecules is of relevance in this context.

3 Exposure and Effect

In the sense of the exposure-effect concept the aniline concentration in the workplace air is defined as the external aniline exposure. Its permissible level is set by the MAK value. The internal exposure is the actual amount of aniline absorbed. It governs the extent to which phenylhydroxylamine and nitrosobenzene are formed metabolically. The effects include the pathophysiological changes in erythrocytes caused by these metabolites. These correlate even after acute intoxication only to a certain extent with the excretion of particular metabolites in the urine. The effects caused by aniline are conventionally evaluated via the changes in the blood methaemoglobin level. This question has been discussed by the Working Group elsewhere (6).

More recently the effects of aniline have been evaluated by measuring the levels of stable protein conjugates. The aniline-haemoglobin conjugate in a blood sample is cleaved hydrolytically and the released aniline determined quantitatively. Bound aniline can be detected even days after exposure. This is possible even in cases in which free aniline and its metabolites have already been completely eliminated. Unlike the determination of methaemoglobin, the results gained by this method are much less dependent on the time of sampling.

4 Selection of the Indicators

There is an extensive publication on this subject (18). Information on exposure and effect relationships on the basis of extensive monitoring of occupationally exposed persons can be found there. Internal exposure can be determined routinely by the measurement of the concentrations of aniline, acetanilide or aminophenols in whole blood or urine. The analysis of aniline should be preferred. The effect can be determined via the level of methaemoglobin or aniline-haemoglobin conjugate.

The importance of the methaemoglobin determination is the possibility it provides of being able to evaluate the extent of acute intoxication quickly. A considerable disadvantage for the setting of a BAT value is the dependency of the findings on the sampling time. The evaluation of the aniline-haemoglobin conjugate is therefore more suitable, in particular for intermittent or chronic exposure.

5 Methodology

Methods to determine the aniline concentration in the blood and urine as well as the analysis of aniline-haemoglobin conjugates have been published in the loose-leaf collection "Analysen in biologischem Material" (Analyses of Hazardous Substances in Biological Materials) (17). This also applies to the determination of methaemoglobin and *p*-aminophenol (2).

6 Evaluation of the BAT Value

Threshold limit values are not established for acute intoxications, which are rare today, but for chronic exposure at the workplace. There are only few investigations on the excretion of aminophenols (9, 26). A tolerance value of 5 mg *p*-aminophenol per l urine was published by Linch (19). Lauwerys (15), on the other hand, recommended permitting internal exposures up to levels causing excretion of 10 mg *p*-aminophenol per l urine. He based his conclusions on investigations in which manifest damage after exposure to aniline was only observed after excretion at *p*-aminophenol concentrations of more than 50 mg/l (20).

In recent reviews (16, 17, 30) the tolerance levels are based almost exclusively on methaemoglobin determinations, seldom on the specific detection of aniline or aminophenol in the body fluids. Generally a methaemoglobin level of 5 % is regarded as tolerable (6).

In accordance with recent results the determination of aniline-haemoglobin conjugates for the monitoring of occupationally exposed persons has gained in importance. In investigations of 1000 employees exposed regularly to aniline and more than 50 employees with acute aniline intoxication it was found that at methaemoglobin levels of less than 5 % also less than 100 µg aniline was released from the haemoglobin conjugate per l whole blood (18). For aniline a BAT value of

100 µg aniline released from the aniline-haemoglobin conjugate/l whole blood

has therefore been set.

When this effect value is observed, damage caused by restriction of the oxygen transport capacity of the erythrocytes can be avoided. In the investigations mentioned above (18) aniline excretion in the urine was also measured. In all cases in which the methaemoglobin level was over 5 % and the aniline level in the haemoglobin conjugate was over 100 µg/l whole blood, aniline excretion was greater than 1 mg/g creatinine. The concentrations of free aniline in the blood were then over 1 mg/l blood. A urine concentration of

1 mg free aniline/l urine

is therefore set as a BAT value for aniline.

7 Interpretation of the Data

After contact with aniline at the level of the MAK value a satisfactory correlation between aniline values in the urine and the aniline-haemoglobin conjugate is only found after three consecutive working days (18). The aniline-haemoglobin conjugate can thus be measured after at least 3 previous working days at any time within a shift period. Also for the determination of free aniline, sampling should take place after at least 3 previous working days.

Simultaneous determination of the blood and urine paramcters dcfined above has been shown to be necessary only at the first examination or immediately after accidents. Routinely the monitoring of persons exposed occupationally to aniline can be limited to the urine investigation. Sampling should be carried out under contamination-free conditions, as far as possible after washing and in private clothes in an occupational medical department.

The aniline concentration in the urine should in the case of high-temperature workplaces be expressed in terms of the creatinine level. For information on the principle of creatinine relativisation reference is made to the special foreword in the loose-leaf collection.

For the intervals between follow-up examinations the “berufsgenossenschaftlicher Grundsatz” G33 (guidelines issued by the Employers' Liability Insurance Association) for occupational-medical health surveillance (11) should be referred to.

8 References

1. Albrecht, W., Neumann, H.-G.: Biomonitoring of aniline and nitrobenzene. Hemoglobin bindings in rats and analysis of adducts. Arch. Toxicol. *57* (1985), 1–5
2. Lewalter, J., Biedermann, P.: Aromatic amines. In: Angerer, J., Schaller, K. H. (eds.): Analyses of Hazardous Substances in Biological Materials, Vol. 4. VCH Verlagsgesellschaft, Weinheim 1994, pp. 67–105
3. Henschler, D. (ed.): Anilin. Gesundheitsschädliche Arbeitsstoffe. Toxikologisch-arbeitsmedizinische Begründung von MAK-Werten, 18th issue. VCH Verlagsgesellschaft, Weinheim 1982
4. Atlas, S. A., Nebert, D. W.: Pharmacogenetics and human disease. In: Parke, D. V., Smith, R. L.: Drug Metabolism from Microbe to Man, Taylor and Francis, London 1976, pp. 393–430
5. Beckett, A. H., Belanger, P. M.: Metabolic N-oxidation of secondary and primary aromatic amines as a route to ring hydroxylation to various N-oxygenated products, and to dealkylation of secondary amines. Biochem. Pharmacol. *25* (1976), 211–214
6. Bolt, H. M., Neumann, H. G., Lewalter, J.: Arbeitsmed. Sozialmed. Präventivmed. *20* (1985), 197–201
7. Cartwright, G. E.: Methemoglobinemia and sulfhemoglobinemia. In: Thorn, G. W. et al. (eds.): Harrison's Principles of Internal Medicine, 8th ed. McGraw-Hill, New York 1977, pp. 1710–1713
8. Damani, L. A.: Oxidation at nitrogen centers. In: Jakoby, W. B., Bend, J. B., Caldwell, J. (eds.): Metabolic Basis of Detoxication. Academic Press, New York 1982, pp. 127–149
9. Dutkiewicz, T., Piotrowski, J.: Experimental investigations on the quantitative estimation of aniline absorption in man. Pure Appl. Chem. *3* (1961), 319–323
10. Evans, E. E, Dinman, A. L., Leech, A. M. A.: Arch. Ind. Health *18* (1958), 422; cited in: Fassett, D. W., Irish, D. D. (eds.): Patty's Industrial Hygiene and Toxicology, 2nd rev. ed. Vol. 2, Toxicology. Wiley & Sons, New York 1958, p. 2110
11. Grundsatz G 33: Aromatische Nitro- und Aminoverbindungen. Berufsgenossenschaftlicher Grundsatz für arbeitsmedizinische Vorsorgeuntersuchungen. Hauptverband der gewerblichen Berufsgenossenschaften e.V. (ed.), 2nd ed. Stuttgart 1981
12. Hanna, P. E., Banks, R. B. in: Anders, M. W. (ed.): Bioactivation of Foreign Compounds. Academic Press, New York 1985, pp. 375–402
13. Irons, R. D., Gross, E. A., White, E. L.: Aniline: evidence for an enterogastric cycle in the rat. Fd. Cosmet. Toxicol. *18* (1980), 393
14. Kao, J., Faulkner, J., Bridges, J. W.: Metabolism of aniline in rats, pigs and sheep. Drug Metab. Disp. *6* (1978), 549
15. Lauwerys, R. R.: Biological criteria for selected industrial toxic chemicals, a review. Scand. J. Work Environ. Health *1* (1975), 139–172
16. Lauwerys, R. R.: Occupational toxicology. In: Doull, J., Klaasen, C. D., Amdur, M. O. (eds.): Toxicology: The Basic Science of Poisons. Macmillan, New York 1980, pp. 699–709
17. Lauwerys, R. R.: Industrial chemical exposure, guidelines for biological monitoring. Biochemical Publication, Davis, California 1983, pp. 97–100
18. Lewalter, J., Korallus, U.: Hemoglobin conjugation and acetylation of aromatic amines – new knowledges for biological monitoring. Int. Arch. Occup. Environ Health. *56* (1985), 179–196
19. Linch, A. L.: Biological monitoring for industrial exposure to cyanogenic aromatic nitro and amino compounds. Am. Ind. Hyg. Assoc. J. *35* (1974), 426–432

19a. Deutsche Forschungsgemeinschaft (DFG): Maximale Arbeitsplatzkonzentrationen und Biologische Arbeitsstofftoleranzwerte 1985. Mitteilung XXI der Senatskommission zur Prüfung gesundheitsschädlicher Arbeitsstoffe. VCH Verlagsgesellschaft, Weinheim 1985

20. Michaux, P., Boiteau, H. L., Tolot, F.: Valeur et limite du depistage clinique et biologique en pathologie professionelle. Arch. mal. prof. med. tra. secur. soc. *32* (1971), 1
21. Nelson, S. D. in: Anders, M. W. (ed.): Bioactivation of foreign compounds. Academic Press, New York 1985, pp. 349–374
22. Neumann, H. G.: Analysis of hemoglobin as a dose monitor for alkylating and arylating agents. Arch. Toxicol. *56* (1984), 1–6
23. Neumann, H. G.: Dosimetry and doseresponse relationships. International seminar on methods of monitoring human exposure to carcinogenic and mutagenic agents. Finland 1983

24. Parke, D. V., Williams, R. T.: Species differences in the *o*- and *p*-hydroxylation of aniline. Biochem. J. *63* (1956), 12
25. Piotrowski, J.: Exposure tests for organic compounds. Industrial Toxicology. US Government Printing Office, Washington D. C. 1977, pp. 70–75
26. Piotrowski, J.: Quantitative estimation of aniline absorption through the skin in man. J. Hyg. Epid. Microbiol. Immunol. *1* (1957), 1–23
27. Smith, J. N., Williams, R. T.: The fate of aniline in the rabbit. Biochem. J. *44* (1959), 242–250
28. Tannenbaum, S. R., Skipper, P. L., Green, L. C., Turesky, R. J., Kadlubar, F. F.: Monitoring exposure to 4-aminobiphenyl via blood protein adducts. International seminar on methods of human exposure to carcinogenic and mutagenic agents. Finland 1983
29. Valentin, H., Lehnert, G., Petry, H., Weber, G., Wittgem, H., Woitowitz, H. J.: Arbeitsmedizin. Band 2: Berufskrankheiten. Thieme Verlag, Stuttgart 1979
30. Valentin, H., Schaller, K. H., Triebig, G. in: Arbeitsmedizin aktuell, Labordiagnostik, Praktische Hinweise für die Durchführung arbeitsmedizinisch-toxikologischer Untersuchungen. G. Fischer Verlag, Stuttgart 1983, pp. 55–70
31. Wieland, E., Neumann, H. G.: Methemoglobin formation and binding to blood constituents as indicators for the formation, availability and reactivity of activated metabolite derived from trans-4-aminostilbene and related aromatic amines. Arch. Toxicol. *40* (1978), 17–35

Author: J. Lewalter
Approved by the Working Group: 10th May 1985

9 Addendum

Evaluation of the aniline concentration in erythrocytes and in the haemoglobin or globin adduct

Metabolism and Kinetics

Aromatic amines can be oxidised during liver metabolism to hydroxylamine compounds and stabilised by conjugation. The glucuronidated aromatic amines are generally eluted via urine. Only small amounts of the conjugates probably penetrate cell membranes such as those of the erythrocytes, and are cleaved there by β-glucuronidase back to the hydroxylamine (1, 2). Some of the phenylhydroxylamine released from various precursors is reduced by hydroxylamine reductase to aniline which can leave the erythrocytes again by diffusion. Some of the hydroxylamine can reach the haem in Hb and be oxidised there to the nitroso compound forming Met-Hb. Only this reactive nitroso intermediate can react with the SH-groups of glutathione (GSH) and globin to yield sulfinamides.

In the blood there is approx. 160 g Hb/l, i. e. approx. 24 mmol SH-groups and approx. 350 mg GSH/l, which corresponds to 1.2 mmol SH-groups. First of all the SH-groups of globin react and then approx. in the ratio of 20 to 1 also those of glutathione (3).

In practice there are two typical forms of exposure:

1. Acute, mainly high aniline exposure
2. Chronic or intermittent (usually low-level) aniline exposure

The two forms of exposure differ in the elimination speed of the products. According to the experience of the Commission (3) transitions are observed between the following extreme types of elimination:

1. Free aromatic amines and aromatic amines released from instable adducts (glutathione conjugates) are excreted with biological half-lives <5 hours.
2. Hb-conjugates are stable and are eliminated during the life cycle of the erythrocytes of approx. three months.

Casuistic Evaluation of the Dominant Processes in Erythrocytes

The consequences of chronic low-level aniline exposures have been observed to date particularly in smokers and those of higher level exposures to aniline particularly after accidental intoxication. In smokers predominantly the accumulated Hb-conjugates of amino aromatics and practically no GSH-conjugates or free aniline are found in the erythrocytes. On the other hand aniline intoxication can lead to erythrocyte concentrations which are degraded in several phases.

After aniline intoxication leading to Met-Hb values of around 40 %, up to 80 % of the aniline in the erythrocytes is eliminated from the erythrocytes in the first 24 hours. A further 10 to 20 % follows in the next 24 hours. Only the last 10 to 20 % of the original erythrocyte aniline can be determined in the erythrocytes in the form of Hb adducts, and that even after a further 10 days (3).

Conclusions and Recommendations

The present BAT value for aniline of 100 μg aniline/l blood released from Hb is based on an evaluation of Hb-adduct formation in several hundred cases of aniline intoxication with Met-Hb values around 5 %. This procedure has proved itself in practice and is recommended for correct occupational-medical surveillance.

Furthermore, for differential diagnosis of unknown levels of aniline comparative evaluation of the total aniline in the erythrocytes and the Hb conjugate formation is recommended. It provides information not only as to the extent of the adduct formation but also as to its cause. Acute aniline intoxication can be recognised by the temporarily increased total aniline levels in the erythrocytes which usually consists of free aniline metabolites and GSH conjugates of nitrosobenzene in addition to considerably smaller amounts of Hb-adducts of aniline. Chronic aniline exposure with Met-Hb values below 5 % or aniline exposure without Met-Hb formation can be recognised above all by the cumulative Hb-adduct formation. In these cases the small amounts of the free aniline metabolites and the GSH-conjugates can no longer be determined with the presently available methods of analysis.

10 References (Addendum)

1. Turesky, R. J., Lang, N. P., Butler, M. A., Teitel, C. H., Kadlubar, F.: Metabolic activation of carcinogenic heterocyclic amines by human liver and colon. Carcinogenesis *12* (1991), 1839–1845
2. Goldstein, J. A., Faletto, M. B.: Advances in mechanisms of activation and deactivation of environmental chemicals. Environ. Health Perspect. *100* (1993), 169–176
3. Lewalter, J.: unpublished investigations (1984–1993)

Author: J. Lewalter
Approved by the Working Group: 27th May 1993

2-Bromo-2-chloro-1,1,1-trifluoroethane (Halothane)

BAT	2.5 mg trifluoroacetic acid/l blood Sampling time: end of the working week, end of exposure
Date of evaluation	1982
Synonyms	Halothane
Formula	F_3C-CHBrCl 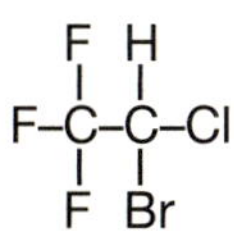
Molecular weight	197.4
Melting point	no data
Boiling point	50.2 °C
Vapour pressure at 20 °C	241.5 mbar
MAK	5 ml/m^3 (ppm) 40 mg/m^3

1 Pharmacokinetics and Metabolism

1.1 Pharmacokinetics

The highly lipophilic 2-bromo-2-chloro-1,1,1-trifluoroethane is quickly absorbed as vapour via the alveolar membrane. As a result of the relatively high blood/gas distribution coefficient of 2.3 on inhalation of unchanging concentrations a constant concentration builds up within relatively short periods of time, which, depending upon the exposure concentration, varies between 20 minutes (narcotic concentration) and several hours (workplace exposure). The high distribution coefficient fat/blood results in accumulation mainly in the fat deposits. Depending on the very different blood supply a steady state condition in the body fat is only attained after many hours, in the lipid of the brain, however, in the course of the first hour. After the end of inhalation the blood concentration decreases by exhalation and metabolism according to first order kinetic functions. Exhalation via the lungs takes place in three phases, whereby the half-time of the first (fast) phase is 20–30 minutes (1), that of the slow (third) phase, however, 2 hours (2).

1.2 Metabolism

Halothane is metabolised rapidly. After uptake of narcotic doses, 15–20 % of the amount of halothane taken up by the body has been shown to be metabolised (3). However, after uptake of lower concentrations such as occurs at the workplace, the proportion metabolised approaches 100 % (4).

The metabolism is effected by mixed function, cytochrome P 450-dependent oxygenases; predominantly in the endoplasmatic reticulum of the liver cell. It can take place via two different routes, the proportions of which are determined by the current oxygen pressure (c. f. Fig. 1): with normal O_2 saturation the oxidative reaction on the C-2 atom is predominant. This yields trifluoroacetic acid; a bromide ion and a chloride ion are eliminated. Which metabolic intermediate is formed has not yet been clarified; to be considered are a free radical formed by elimination of bromine and a hydroxyl compound produced by the insertion of an O atom; the latter would lose HBr spontaneously to yield trifluoroacetyl chloride which would be transformed by hydrolysis to free trifluoroacetic acid.

Under conditions of hypoxia which can occur in vivo, elimination of hydrogen fluoride takes place forming 1,1-difluoro-2-bromo-2-chloroethene, a very reactive compound. This can react with macromolecules forming covalent bonds, possibly being oxidised itself to an even more reactive epoxide or coupling with glutathione to yield a mercapturic acid derivative by known metabolic pathways. Trifluoroacetic acid ethanol amide has been identified as a further metabolite (5) whose mode of formation could not be clarified definitely; it occurs here, however, only in traces.

1.3 Pharmacokinetics of Trifluoroacetic Acid (TFA)

Trifluoroacetic acid is the main metabolite. It is the most suitable of all the metabolites for biological monitoring: it can be unambiguously determined (6), is excreted in the urine as almost the only metabolite and is formed during exposure to concentrations which occur at the workplace in practically invariable proportions (7).

The half-life of trifluoroacetic acid has been determined in experiments in humans with a ^{14}C-labelled compound (7). In two individuals after i. v. injection it was 66 hours in both cases (7). After 6 hours inhalation of 20 ppm halothane in one test person it came to 52 hours, after inhalation of 100 ppm for the same length of time in another person to 60 hours (7). In two patients who had received a two hour continuous narcosis with approx. 7500 ppm halothane 48 hours and 52 hours were measured. In all cases the kinetic functions were strictly first order (7).

On administration of ^{14}C-labelled trifluoroacetic acid to two test persons the distribution volume was determined as 21 % and 22 % of the body weight, respectively.

Binding to the erythrocytes was slight (4–14 % depending upon the concentration). On the other hand the plasma protein binding (49–54 % at concentrations of 1–128 μg/ml) is relatively high. 99 % is excreted in the urine, only 1 % in the faeces (7).

2 Relationship of 2-Bromo-2-chloro-1,1,1-trifluoroethane in Ambient Air to Trifluoroacetic Acid in Biological Material

In a systematic study the relationship of low halothane concentrations inhaled long-term to the formation of trifluoroacetic acid levels in the blood was investigated (7). Six test persons each inhaled daily for 6 hours on five consecutive days analytically controlled concentrations of 8.2 ppm, 21.7 ppm and 49.3 ppm halothane (average values with little scatter). The trifluoroacetic acid levels were determined by gas chromatography (8) before and after the 6-hour exposures and 2 hours after the end of exposure; in addition the trifluoroacetic acid concentration in the 12-hour urine was determined. In the week following exposure once a day the blood level as well as the concentration in the 24-hour urine was determined (7).

There was an increase in the blood TFA concentration at all three halothane concentrations during the 6-hour exposure period. In the intervals between exposures it decreased regularly. Towards the end of the exposure week an equilibrium concentration in the blood developed. The elimination half-times were 40–60 hours, fluctuating somewhat from study to study. In a study over 2 weeks with 22 ppm halothane in the inhaled air the concentration of TFA in the blood in the second week did not increase above the level in the first week.

There is a clear correlation between the TFA concentrations in blood and urine measured simultaneously (Fig. 2), and it remains constant over a range of 0.2–9 μg/ml TFA in blood.

There is a linear relationship between the halothane concentration constantly breathed in during the 6-hour inhalation period and the accumulated trifluoroacetic acid concentration in the blood at the end of the exposure week (Fig. 3). The linear character of the relationship allows direct extrapolation to an 8-hour exposure interval. According to this, an inhaled concentration of 10 ppm halothane corresponds to a TFA blood concentration of 5 μg/ml (7). This relationship was also found in investigations at the workplace where the halothane levels were determined by air analysis (9).

3 Critical Toxicity

After long-term exposure at workplace concentrations two kinds of damage are attributed to halothane: liver diseases and embryotoxic effects. Both can, despite several comprehensive field studies, only be regarded as suppositions; definite confirmation and experimental substantiation are lacking. Reference is made to the extensive review of halothane in “Toxikologisch-arbeitsmedizinische Begründung von MAK-Werten” (occupational-medical toxicological documentation of MAK values) (10).

Trifluoroacetic acid is a very strong organic acid. Its acute toxicity is very low (11). There are no relevant investigations of chronic toxic effects (yet). Its plasma protein binding of 50 % is in some cases high enough to displace certain drugs from the same binding sites and thus to produce relative overdoses of these medicines. Yet there are equally few relevant investigations of this effect.

It is possible that the highly reactive metabolic intermediates produced from halothane are responsible for the organotoxic effects: trifluoroacetyl chloride, trifluorochloroethane radicals and difluorochlorobromoethane as well as possibly its epoxide. The latter compounds are formed increasingly with decreased oxygen pressure. Under such conditions the increased cleavage of fluoride is also measured (12). However, the end products of this metabolic pathway are found in such low quantities, even under conditions of severe oxygen deficiency, that they cannot be used for biological monitoring.

4 Selection of the Indicators

Fundamentally several indicators are available for analysis: halothane in the exhaled air, halothane in blood, trifluoroacetic acid in blood and trifluoroacetic acid in urine. The halothane determination in exhaled air and blood is relatively simple and sensitive methods are available (13). However, at workplaces greatly fluctuating concentrations and irregular interruptions of the exposures must be expected. Due to the short half-times of the first two phases of elimination frequent measurements have to be carried out to evaluate the total exposure as the concentrations at the end of a work shift or a working week cannot reflect the total uptake. Therefore practical difficulties make this method of determination unsatisfactory.

The correlation between trifluoroacetic acid concentrations in the blood and urine is satisfactory for practical purposes. It presupposes the collection of 12-hour urine (c. f. Fig. 2), with shorter collection periods the scatter is expected to be much greater. In order to determine the exposure during a week more frequent determinations, preferably daily, would have to be carried out. The determination of TFA in urine is thus a method of biological monitoring which is practicable in principle.

Easier and at the same time reflecting the week's exposure more precisely is, however, trifluoroacetic acid determination in blood at the end of an exposure week. Due to the long half-life of trifluoroacetic acid and the resulting continuous accumulation, greatly fluctuating exposures result in only slight deviations from the actual (total) average value of the week's exposure which can be disregarded for practical purposes. The determination of the blood trifluoroacetic acid concentration at the end of the working week is the preferred method.

5 Evaluation of the BAT Value

According to extensive reports in the literature, long-term exposure to low concentrations of halothane in the operating theatre is supposed to have embryotoxic, possibly mutational effects. Definite proof of this has, however, not yet been produced. Neither can threshold limit values for the above effects be derived from animal experiments (10). Risk evaluation and control must therefore be based on hypotheses. The chosen parameter, the formation of a strongly cumulative metabolite, trifluoroacetic acid, is that used for the establishment of the MAK value. The accumulation of trifluoroacetic acid should be kept within safe limits. It is to be expected that trifluoroacetic acid – like trichloroacetic acid, a metabolite of trichloroethene – competes with a series of frequently used medicines for plasma protein binding (14, 15) and so increases the effects of these drugs. Furthermore, halothane interferes with the biotransformation of drugs by the degrading enzymes of the liver cell; this interference can also cause increased effects of drugs taken at the same time. It is therefore suggested that the following BAT value is used provisionally:

2.5 mg trifluoroacetic acid/l blood

This corresponds to the currently valid MAK value of 5 ppm halothane in the ambient air as an 8-hour average value during a working week.

6 Interpretation of the Data

The BAT value should be determined at the end of the last shift of a week. It should be taken into account that for comparison of the halothane concentration in the ambient air and the trifluoroacetic acid concentration in whole blood, average values have been used. The BAT value is thus not, as is usually the case, regarded as an individual peak value. With regard to the provisional nature of the BAT value and the impossibility at present of determining threshold limit values for the critical toxicity, however, this procedure seems justified.

The determination of trifluoroacetic acid in urine can also be used as the starting point for the evaluation of workplaces. A concentration of 10 mg/l trifluoroacetic acid in urine then corresponds to a trifluoroacetic acid concentration in blood of 2.5 mg/l. For this determination a 12-hour urine sample or at least an 8-hour urine sample should be collected during the last shift of a working week.

7 References

1. Gostomzyk, J. G., Eisele, G., Bardua, F.: Z. prakt. Anaesth. *7* (1972), 245
2. Dallmeier, E., Henschler, D.: Unpublished results (1978). Institut für Pharmakol. und Toxikol., Versbacher Str. 9, Würzburg

3. Stier, A., Alter, H., Hessler, O., Rehder, K.: Anesth. et Analg. *43* (1964), 723
4. Rösch, J.: Med. Inauguraldissertat., Würzburg 1981
5. Cohen, E. N., Trudell, J. R., Edmunds, H. N., Watson, E.: Anesthesiology *43* (1975), 392
6. Dallmeier, E.: 1,1,2-Trichlor-1,2,2-trifluorethan in Serum. In: Henschler, D. (ed.): Analytische Methoden zur Prüfung gesundheitsschädlicher Arbeitsstoffe. Band 2: Analysen in biologischem Material. 6th issue VCH Verlagsgesellschaft, Weinheim 1982
7. Dallmeier, E., Henschler, D.: Dtsch. med. Wochenschr. *106* (1981), 324
8. Witte, L., Nau, H., Fuhrhop, J. H., Doenicke, A., Grote, B.: J. Chromatogr. *143* (1977), 329
9. Bencsath, F. A., Weichardt, H.: Massenspektrometrische Identifizierung und automatisiertes gaschromatographisches Monitoring flüchtiger Anästhetika in der Luft und im Blut. Berichte über die 18. Jahrestagung der Deutschen Gesellschaft für Arbeitsmedizin. Gentner, Stuttgart 1978, p. 93
10. Henschler, D. (ed.): Halothan. In: Gesundheitsschädliche Arbeitsstoffe. Toxikologisch-arbeitsmedizinische Begründungen von MAK-Werten. 7th issue. VCH Verlagsgesellschaft, Weinheim 1979
11. Airaksinen, M. M., Tammisto, T.: Ann. Med. exp. Fenn. *46* (1968), 242
12. Van Dyke, R. A., Gandolfi, A. J.: Drug. Metab. Dispos. *4* (1976), 40
13. Schaller, K. H., Goßler, K.: Halothan. In: Henschler, D. (ed.): Analytische Methoden zur Prüfung gesundheitsschädlicher Arbeitsstoffe. Band 2: Analysen in biologischem Material. 3rd issue. VCH Verlagsgesellschaft, Weinheim 1978
14. Müller, G., Spassovski, M., Henschler, D.: Arch. Toxicol. *29* (1972), 335
15. May, R.: Dissert. Universität Würzburg 1977

Author: D. Henschler
Approved by the Working Group: 14. 6. 1982

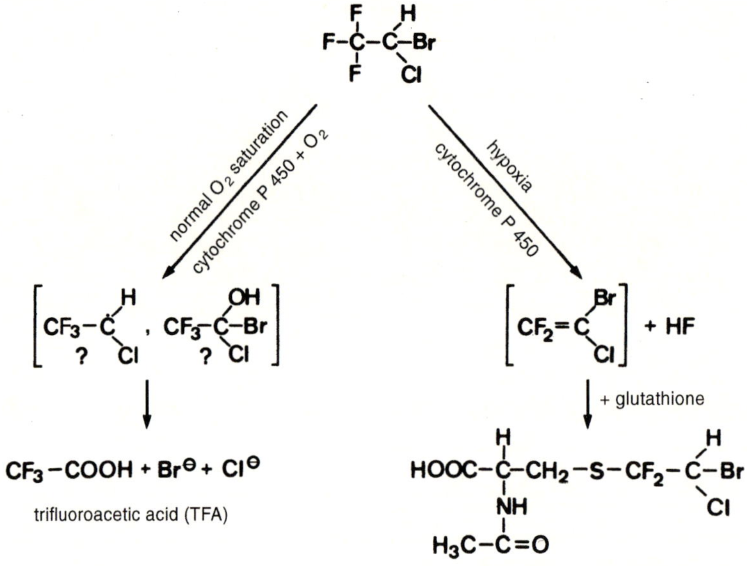

Fig. 1: Diagram of the pathways of halothane metabolism and influence of the current oxygen concentration in the liver

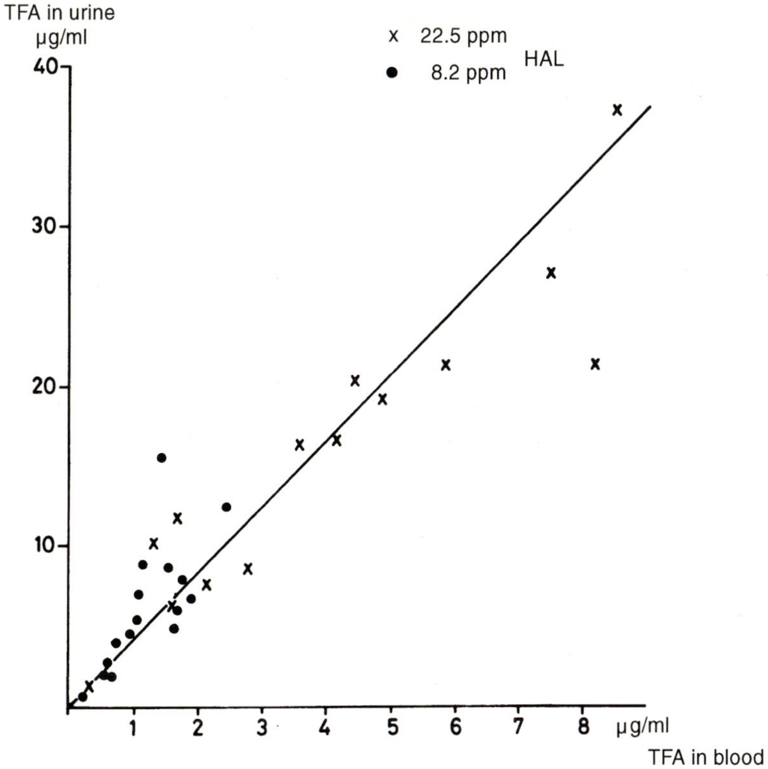

Fig. 2: Daily 6-hour exposures of two different groups of test persons (n = 6) with two different halothane concentrations over 5 days. Ordinate (linear): urine trifluoroacetic acid (TFA) concentration. Abscissa (linear): blood trifluoroacetic acid concentration during the urine sampling period

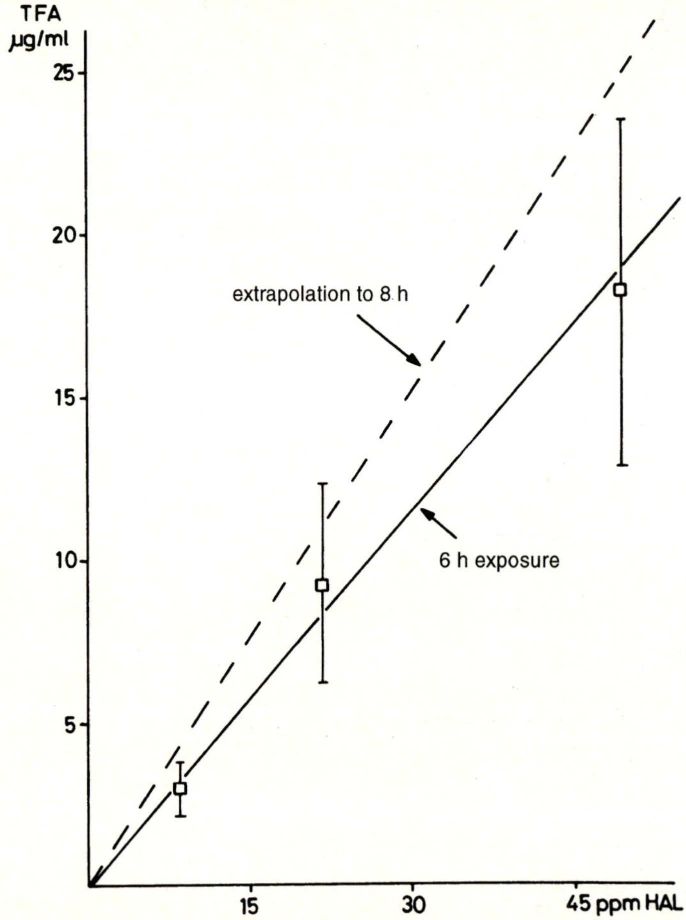

Fig. 3: Correlation of the maximum blood trifluoroacetic acid (TFA) concentration with the halothane concentration during daily 6-hour inhalation on 5 weekdays. Extrapolation to an 8 h work shift

2-Butanone

BAT	5 mg 2-butanone/l urine Sampling time: end of exposure
Date of evaluation	1988
Synonyms	Methyl ethyl ketone MEK Ethyl methyl ketone
Formula	$CH_3–CO–C_2H_5$
Molecular weight	72.11
Melting point	−85.9 °C
Boiling point	79.6 °C
Vapour pressure at 20 °C	105 hPa
MAK	200 ml/m^3 (ppm) 590 mg/m^3

1 Metabolism and Kinetics

At the workplace 2-butanone (methyl ethyl ketone, MEK) is taken up predominantly via inhalation. The ketone is also able to enter the body in quite considerable amounts via the intact skin.

The metabolism of 2-butanone in the human body is only known in part (Fig. 1). Approx. 25 % of the ketone taken up by the body is eliminated unchanged via exhalation (1). The rest of the ketone is metabolised. Some is oxidised to 3-hydroxy-2-butanone which could be identified in the urine of exposed workers. Part of the 2-butanone absorbed is excreted unchanged in the urine. Together with the 3-hydroxy-2-butanone the amount of 2-butanone excreted in the urine accounts for only approx. 0.1 % of the dose taken in (2). In experimental animals, apart from the 3-hydroxy-2-butanone, 2,3-butanediol and 2-butanol were identified as products of 2-butanone metabolism (3).

It can be assumed that these metabolites are also formed in the human organism and then excreted in the urine. The hydroxy compounds are most probably partially bound to sulphates or glucuronic acid. The relation between the amounts of the various metabolites is at present unknown (Fig. 1).

2-Butanone, as a fairly readily water-soluble solvent, is only eliminated to a relatively small extent, as already described, via exhalation. During the examination of persons occupationally exposed to 2-butanone it was found that the concentration of the toxic substance in the alveolar air (c_a) was approx. a quarter of the ambient air concentration (c_i) ($c_a/c_i = 0.27$) (4).

2-Butanone accumulates in blood at higher concentrations than in the ambient air. Persons occupationally exposed to 2-butanone show concentrations of the toxic substance in the blood during exposure which according to Perbellini et al. (2) are approx. 35 times higher than those in the ambient air (c_b/c_i).

According to this study the 2-butanone concentration in urine is lower. The ratio of the 2-butanone concentration in urine to that of ambient air (c_u/c_i) was determined as 1/4.8.

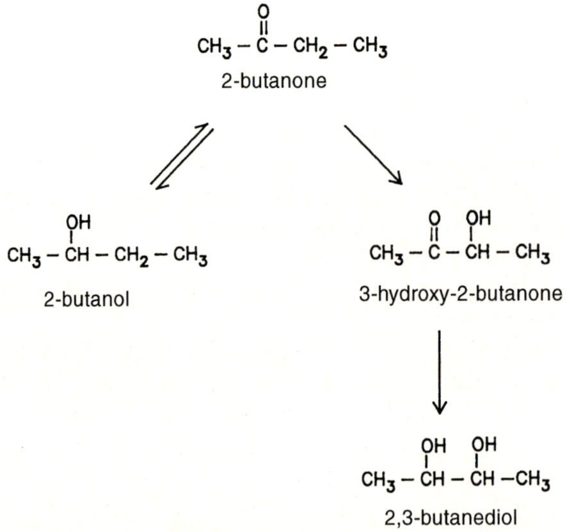

Fig.1: Metabolism of 2-butanone (3).

A close relationship has been found under workplace conditions between the 2-butanone concentration in ambient air on the one hand and the solvent concentration in alveolar air and in urine on the other (2). These authors did not reveal a correlation between the concentration of the toxic substance in blood and that in urine.

According to a mathematical model the 2-butanone concentration in the blood increases during the whole exposure period and after eight hours reaches approx. 16 % of the saturation value. After the end of exposure the 2-butanone is eliminated from the blood with a half-life of 270 min. 16 hours after the end of exposure the blood level has decreased to an insignificantly small value so that one can simply assume that after occupational exposure there is no accumulation of 2-butanone in the body.

The excretion of 2-butanone in urine follows the ketone level in the blood. Four hours after the start of exposure the butanone concentration in urine reaches values which even after continuing exposure increase only insignificantly.

An equilibrium between the ambient air concentration and the blood concentration of 2-butanone is reached much more slowly with 2-butanone than with less hydrophilic solvents such as e.g. aromatic hydrocarbons and to an even greater extent aliphatic hydrocarbons. This is probably mainly due to the fact that 2-butanone distributes itself relatively equally over the various compartments of the body. This can be seen from the distribution coefficients tissue/air which are very similar for the various tissues (2).

2 Critical Toxicity (see also MAK documentation)

Although 2-butanone is one of the most frequently used solvents only little is known about the systemic effects of this ketone on man. This is due not least to the fact that the irritating effects of the toxic substance on the eyes and mucous membranes prevent the uptake of higher doses at the workplace.

The odour threshold of 2-butanone is about 25 ppm (5). Concentrations in the ambient air of as little as 100 ppm can lead to slight irritation of the eyes. 2-Butanone concentrations above the currently valid MAK value of 200 ppm cause marked irritation of the eyes as well as mucous membranes, nose and throat (6).

With ambient air concentrations of more than 300 ppm the central depressive effects of 2-butanone come to the fore. Furthermore, with such concentrations headache (7) and "stomach ache" (8) were reported.

Whether 2-butanone also has neurotoxic properties cannot be clarified definitively at present. It can be regarded as certain that 2-butanone potentiates the neurotoxic effects of *n*-hexane and its metabolites 2-hexanol, 2-hexanone and 2,5-hexanedione. Spencer et al. (9) report about 79 cases of sensomotor polyneuritis in a factory where a mixture of 90 % 2-butanone and 10 % 2-hexanone (methyl-*n*-butyl ketone) was used as a solvent. In another case there was also an increase in the occurrence of sensomotor polyneuritis with the use of a 2-butanone/2-hexanone mixture (10). Here it should be particularly noted that the 2-butanone concentration in the ambient air was very low with values between 6 and 36 ppm. If 2-hexanone is present alone in the ambient air then about ten times higher concentrations are required to produce comparable effects.

The intensifying of the specific neurotoxic (peripheral) effect of *n*-hexane through 2-butanone has been studied in animal experiments. Two Japanese groups (11, 12) found greater changes in the neurophysiological parameters after simultaneous application of hexane and 2-butanone than were observed after exposure to pure hexane. This combination effect is apparently not due to metabolic interference, as Iwata et al. (13, 14) discovered no increase but rather a decrease in the 2,5-hexanedione excretion in the urine after simultaneous exposure to 2-butanone and *n*-hexane. 2,5-Hexanedione is regarded as the ultimate neurotoxic metabolite of *n*-hexane.

After simultaneous exposure to 2-butanone and toluene no additive effects on the CNS were discovered in an experimental study on volunteers. The 2-butanone concentration was 200 ppm, the toluene concentration 100 ppm (15).

3 Selection of the Indicators

At present especially the following parameters may be considered for the biological monitoring of persons exposed to 2-butanone:

- 2-butanone concentration in urine
- 2-butanone concentration in blood
- 2-butanone concentration in the alveolar air
- 3-hydroxy-2-butanone excretion in urine.

2-Butanone is sufficiently hydrophilic to enter the kidneys and be excreted in the urine. As has been shown in various studies, there is, therefore, also a close relationship between the 2-butanone concentration in the air at the workplace and the 2-butanone excretion in urine (see section 1). This proportionality makes the 2-butanone concentration in urine fundamentally a suitable parameter for the estimation of internal exposure. A further argument in favour of this parameter is that the irritant effects on the mucous membrane and also the central nervous effects of 2-butanone, i.e. the main toxic principle, are produced by the unchanged molecule of the toxic substance.

Practical and also analytical and methodological reasons speak in favour of the 2-butanone concentration in the urine as a parameter for the estimation of internal exposure. The determination of 2-butanone in urine is analytically and diagnostically specific which has led to there being sufficient data for the 2-butanone concentration in urine for the evaluation of a BAT value. A further advantage of this parameter is that the concentration of the toxic substance in urine reflects, as it were, the integral of the dose during the whole duration of exposure. As a disadvantage of this parameter the varying dilution of the urine samples and its influence on the 2-butanone concentration must be accepted.

Diagnostically even more sensitive and, unlike urine, the 2-butanone concentration in blood is not dependent upon the activity of the kidneys. As can be indirectly inferred from the results of Perbellini et al. (2) at steady state the 2-butanone concentration in the blood is approx. 7 times as high as that in urine. During investigation by members of the Commission of various collectives exposed to 2-butanone different results were found for the distribution of the ketone between blood and urine (Fig. 2). According to this investigation the relationship between the 2-butanone concentration in blood (c_b) and in urine (c_u) proved to be dependent upon the collective. The ratio c_b/c_u was between 0.5 and 2.5. A seven-fold higher concentration in the blood level than in the urine was not found even in a single case. From these results it can be concluded that the absorption and excretion kinetics of 2-butanone differ greatly in blood and urine. This means that the relationship of the ketone concentration in blood to that in urine depends markedly on the profile of the changes in the concentration of the 2-butanone in the air with time and on the time of taking the sample of biological material. It may be assumed that the steady state concentration of ketone in the blood is at least 2.5 times greater than that in urine. This was the largest value found in this investigation. Although it would be desirable in principle to define a BAT value for the 2-butanone concentration in blood the database described is not sufficient.

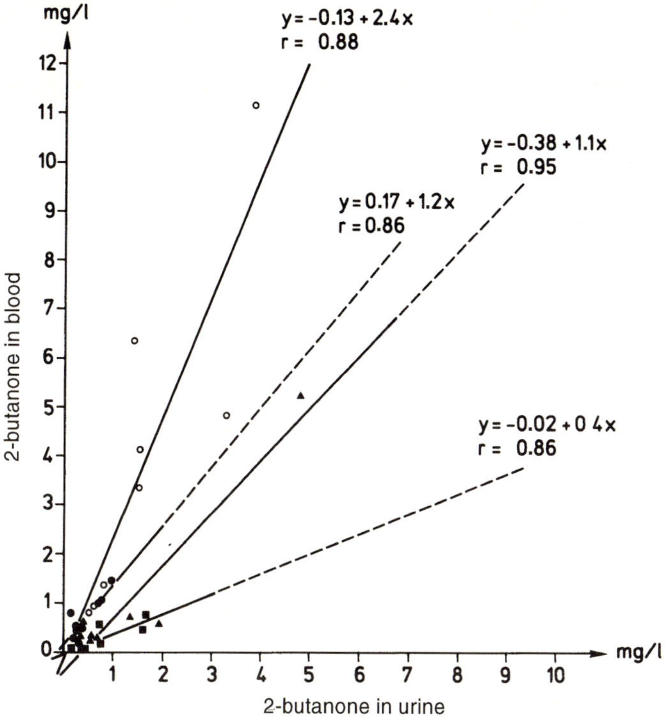

Fig. 2: Relationships between the concentrations of 2-butanone in the blood and urine under various conditions of exposure.

Fundamentally, the concentration of organic solvents in alveolar air represents a diagnostically specific and sensitive way of objectively determining internal exposure. The determination of this parameter is, however, accompanied by a series of problems. Among these are the difficulties and intrusiveness of taking samples of expired air. The transport and storage of alveolar air samples can also lead to errors in the measured values.

The disadvantage of 3-hydroxy-2-butanone in the urine as a parameter for the estimation of an internal 2-butanone exposure is that this metabolite is only formed in relatively small amounts from 2-butanone. The diagnostic sensitivity of this parameter is therefore not sufficient for health surveillance when monitoring persons with low level exposure. 3-Hydroxy-2-butanone is also said to be taken up with a normal diet. 3-Hydroxy-2-butanone excretion in urine therefore does not necessarily show an external occupational exposure to 2-butanone, which limits the diagnostic specificity of this parameter.

4 Methodology

The determination of 2-butanone in urine can be carried out simply and reliably by means of gas chromatographic head-space analysis (16). In particular the use of a fused silica capillary column with a chemically bound phase allows this head-space procedure to be used extremely selectively (17, 18).

The working group “Analytische Chemie“ (“Analytical Chemistry“) of the DFG Commission for the Investigation of Health Hazards of Chemical Compounds in the Work Area has developed a capillary gas chromatographic head-space procedure that allows the aliphatic ketones used as industrial solvents to be determined in biological matrices in a single analytical procedure.

5 Evaluation of the BAT Value

The evaluation of a BAT value for the 2-butanone excretion in urine is possible via the relationship between external exposure, internal exposure and effect. The present MAK value is 200 ppm (19). Central nervous effects occur first with still higher ambient air concentrations of 300 ppm or more. After exposure to concentrations below 200 ppm (corresponding to 590 mg/m^3) so far no detrimental effects of 2-butanone on the health of the employees have been observed.

With ambient air concentrations which meet the requirements of present-day occupational hygiene a significant correlation has been found between the external 2-butanone exposure and the excretion of 2-butanone in urine. On the basis of this relationship a BAT value can be established for this parameter. Table 1 contains the conversion factors known from the literature for external and internal 2-butanone exposure. With ambient air concentrations of 200 ppm 2-butanone concentrations in urine between 2.1 mg/l (2) and 5.3 mg/l (16) are to be expected. For the evaluation of a BAT value valid for the individual it must be noted that these values represent average values for collectives exposed to an ambient air concentration at the level of the threshold limit values. The results of members of the Commission on various collectives exposed to 2-butanone (Tab. 1) confirm the results of Perbellini et al. (2). For the 2-butanone concentration in the urine a BAT value of

5 mg 2-butanone per litre urine

has therefore been set.

A BAT value for the 2-butanone concentration of blood can at present not be established. The data available so far are not sufficient. The evaluation of such a value would seem, however, to be desirable.

Table 1: Relationship between the external exposure to 2-butanone and the excretion in urine

Collective and occupation	2-Butanone in air mg/m^3 $\bar{x} \pm s$ (range)	2-Butanone in urine µg/l $\bar{x} \pm s$ (range)	x2-butanone(a)-y2-butanone(u)	Authors
a) Published data				
27; shoe factory	101 ± 67 (8–272)	487 ± 277 (120–1120)	y = 3.2x + 196*	Perbellini et al. 1984 (2)
34; various industries	(up to 1000)		y = 3.2x + 320*	Ghittori et al. 1987 (20)
62; ♂ printing workers	(up to 300)		y = 26.3x + 53**	Miyasaka et al. 1982 (16)
b) Investigations by Commission members (in shortened form published by Angerer et al. (18))				
10; ♂ mechanics without respirators	12.4 ± 5 (5–21)	177 ± 64 (89–247)		
9; ♂ paint sprayers with respirators	161 ± 34 (16–210)	1249 ± 1451 (304–1936)		
8; ♂ paint sprayers with respirators	85 ± 136 (16–422)	805 ± 609 (146–1682)		
9; ♂ paint sprayers with respirators	84 ± 157 (22–472)	441 ± 299 (146–978)		

* x in µg/l; ** x in ppm

6 Interpretation of the Data

The urine samples for the determination of the 2-butanone concentration should be collected at the end of exposure. Immediately after urine collection, aliquots, as a rule 2 ml, should be pipetted into head-space vials (20 ml) which should then be sealed immediately with a gas-tight Teflon-coated rubber stopper. If this portioning into gas-tight head-space vials does not take place immediately and if the urine is left standing for longer periods after sampling at room temperature then, due to evaporation of 2-butanone, losses are to be expected. The samples can be transported and stored in these head-space vials. According to data available today, three-day storage at room temperature does not lead to errors. Storage of the urine samples for longer periods should take place in the deep-freezer at −18 °C.

The 2-butanone concentration in urine should be stated and evaluated as mass per volume, i.e. in mg/l. Relation to creatinine is not appropriate (16). Apparently 2-butanone excretion follows other mechanisms than those for creatinine. According to data from the same authors a relation of the mass of the excreted creatinine to the specific weight has not proved useful. According to the general procedure of the Commission (see Occupational-Medical Toxicological Documentation of BAT Values, special foreword: a) Creatinine as a Parameter in the Evaluation of the Concentration of Substances in Urine) the 2-butanone concentration should only be determined if the urine being analysed has a specific weight >1.010.

As already mentioned 2-butanone potentiates the neurotoxic effects of 2-hexanone and *n*-hexane. If *n*-hexane, 2-hexanol or 2-hexanone occur at the workplace together with 2-butanone then the BAT value established here for 2-butanone does not apply. If in such cases neither 2-butanone nor *n*-hexane (or 2-hexanol/2-hexanone) can be replaced at the workplace then, for practical reasons, according to the literature references quoted in the above documentation it can be assumed that below 1/10 of the BAT value mentioned here, even under conditions of mixed exposure, no peripheral neurotoxic effects are to be expected.

7 References

1. Tada, O., Nakaaki, K., Fukabori, S.: Rodo Kagaky *48* (1972), 305–310
2. Perbellini, L., Brugnone, F., Mozzo, P., Cocheo, V., Caretta, D.: Int. Arch. Occup. Environ. Health *54* (1984), 73–81
3. DiVicenzo, G. D., Kaplan, C. J., Dedinas, J.: Toxicol. Appl. Pharmacol. *36* (1976), 511–522
4. Brugnone, F., Perbellini, E., Gaffuri, E., Apostoli, P.: Int. Arch. Occup. Environ. Health *47* (1980), 245–261
5. Shell Chemical Corp.: Ind. Hyg. Bull. Toxicity Data Sheet, Methyl ethyl ketone, SC 57–109, 1959, quoted in: Clayton, G. D., Clayton, F. E. (eds.): Patty's Industrial Hygiene and Toxicology, 2nd. rev. Vol 2C, Toxicology. Wiley & Sons, New York 1962
6. Nelson, K., Ege jr., J. F., Ross, M., Woodman, L. E., Silverman, L.: J. Ind. Hyg. Toxicol. *25* (1943), 282
7. Elkins, H. B.: The Chemistry of Industrial Toxicology, 2nd ed. Wiley & Sons, New York 1962
8. Smith, A. R., Mayers, M. R.: Ind. Bull. N. Y. State Dept. Labor *23* (1944), 174; quoted in: Clayton, G. D., Clayton, F. E. (eds.): Patty's Industrial Hygiene and Toxicology, 2nd rev. ed. Vol. 2C. Toxicology. Wiley & Sons, New York 1962, p. 1733
9. Spencer, P. S., Schaumburg, H. H., Raleigh, R. L., Terhaar, C. H.: Arch. Neurol. (Chic.) *32* (1975), 219
10. Billmaier, D., Yee, H. T., Allen, N., Craft, B., Williams, N., Epstein, S., Fontaine, R.: J. Occup. Med. *16* (1974), 665
11. Misumi, J., Nagano, M.: Br. J. Ind. Med. *42* (1985), 155–161
12. Takeuchi, Y., Ono, Y., Hisagana, N., Iwata, M., Aoyama, M., Kitoh, J., Sugiura, Y.: Br. J. Ind. Med. *40* (1983), 199–203
13. Iwata, M., Takeuchi, Y., Hisagana, N., Ono, Y.: Int. Arch. Occup. Environ. Health *53* (1983), 1–8
14. Iwata, M., Takeuchi, Y., Hisagana, N., Ono, Y.: Int. Arch. Occup. Environ. Health *54* (1984), 273–281
15. Dick, R. B., Setzer, J. V., Wait, R., Hayden, M. B., Taylor, B. J., Tolos, B., Putz-Anderson, V.: Int. Arch. Occup. Environ. Health *54* (1984), 91–109
16. Miyasaka, M., Kumai, M., Koizumi, A., Watanabe, T., Kurasako, R., Sato, K., Ikeda, M.: Int. Arch. Occup. Environ. Health *50* (1982), 131–137
17. Kassebart, V., Angerer, J., Heinrich, R., Behling, K., Lehnert, G.: Kombinierte Lösungsmittelexposition bei Lackierarbeiten. In: Bolt, H. M., Piekarski, C., Rutenfranz, J. (eds.): Verhandlungen der Deutschen Gesellschaft für Arbeitsmedizin e. V., 25. Jahrestagung in Dortmund 1985. Gentner Verlag, Stuttgart 1985, pp. 449–453
18. Angerer, J., Kassebart, V., Wolf, D., Behling, K., Lehnert, G.: Kombinierte Lösungsmittelexpositionen bei Beizarbeiten. In: Bolt, H. M., Piekarski, C., Rutenfranz, J. (eds.): Verhandlungen der Deutschen Gesellschaft für Arbeitsmedizin e. V., 25. Jahrestagung in Dortmund 1985. Gentner Verlag, Stuttgart 1985, pp. 461–465
19. Deutsche Forschungsgemeinschaft: Maximale Arbeitsplatzkonzentrationen und Biologische Arbeitsstofftoleranzwerte. Mitteilung XXVI der Senatskommission zur Prüfung gesundheitsschädlicher Arbeitsstoffe. VCH Verlagsgesellschaft, Weinheim 1990
20. Ghittori, S., Imbriani, M., Pezzagno, G., Capodaglio, E.: Am. Ind. Hyg. Assoc. *48* (1987), 786–790

Author: J. Angerer
Approved by the Working Group: 15th March 1988

N,N-Dimethylformamide

BAT	15 mg *N*-methylformamide/l urine[1)] Sampling time: end of shift 1) Sum of *N*-methylformamide and *N*-methyl-*N*-hydroxy-methylformamide with gas chromatographic determination
Date of evaluation	1993
Synonyms	*N*-Formyldimethylamine DMF
Formula	C_3H_7NO

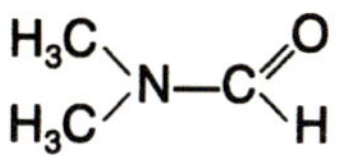

Molecular weight	73.10
Melting point	−61 °C
Vapour pressure at 20 °C	3.5 hPa
Solubility in water	miscible in all proportions
MAK [last established: 1992]	10 ml/m^3 (ppm) 30 mg/m^3

1 Metabolism and Kinetics

Dimethylformamide enters the human organism through inhalation of the vapour and through direct skin contact (8, 12). Under steady-state conditions 65-83 % of the DMF taken up via inhalation is retained by the alveoli and absorbed (2). In an experimental study on volunteers absorption as high as 90 % was reported (19).

On direct skin contact with the liquid, DMF is absorbed very rapidly. In humans the penetration rate is 9 mg/cm^2/h after a delay of 1 to 1 1/2 hours (18). After submersion of a hand in liquid DMF for 15 minutes renal DMF metabolite excretion is twice as high as after eight-hour inhalation of 10 ppm DMF (18).

Gastro-intestinal uptake is of minor importance at the workplace.

1.1 Physiological Metabolism and Kinetics

Absorption of DMF via the skin from the vapour phase has been described (19). This is also possible from contaminated working clothes (7). After occupational exposure skin absorption therefore plays an important role. DMF is one of the substances for which TRGS 150 (Technical Guidance Note for Hazardous Chemicals) stipulates the monitoring of exposure to the toxic substance in biological material.

Biotransformation takes place in the liver; DMF is metabolised rapidly. Only after very high-level exposure to DMF is a small amount of the absorbed dose excreted unchanged in the urine and via the gastro-intestinal tract (9). In animal experimental studies it was shown that after exposure to high concentrations of DMF the metabolism becomes saturated and that very high doses of DMF can inhibit its own metabolism.

As studies on humans and animal experiments show, the main product of the biotransformation of DMF, *N*-hydroxymethyl-*N*-methylformamide (DMF-OH), is formed by enzymatic oxidation. Demethylation to *N*-hydroxymethylformamide and *N*-methylformamide (NMF) follows. By further demethylation formamide is formed (F). This metabolic pathway is shown schematically in figure 1. The metabolites are excreted via the urine; free NMF and F are only found in low concentrations. Unmetabolised dimethylformamide (DMF) is also excreted renally, albeit in extremely small amounts. A further metabolite is the mercapturic acid, *N*-acetyl-*S*-(*N*-methylcarbamoyl)cysteine (AMCC). The excretion of this mercapturic acid in urine accounts for approx. 14 % of the DMF taken up.

The elimination of *N*-methylformamide reaches its maximum at the end of exposure and the substance is then eliminated with biological half-times of three to seven hours. In the literature both monophasic and also biphasic kinetics are described. In a recent study with volunteers, the average half-life of *N*-methylformamide was four hours after exposure to DMF levels of 10 to 60 mg/m^3 (19).

As a consequence of this rapid elimination, *N*-methylformamide itself does not accumulate in the organism.

On the other hand AMCC is excreted relatively slowly. Half-lives between 20 and 30 hours have been described (on average 23 hours); accumulation has also been reported (19).

These values apply for uptake of DMF both by inhalation and through the skin.

1.2 Transformation during Analysis

The DMF metabolites (DMF-OH, NMF-OH, NMF, F) are as a rule determined by gas chromatography. It should be taken into account that by the thermal cleavage of the hydroxymethyl group DMF-OH and NMF-OH are determined as the corresponding NMF and F compounds (see Fig. 1). The cleavage of the hydroxymethyl group is complete at a temperature of 250 °C in the injector block of the gas chromatograph. The NMF determined gas chromatographically thus corresponds to the sum of DMF-OH and the normal NMF concentration excreted. Similarly the determined formamide is made up of the sum of NMF-OH and the eliminated formamide.

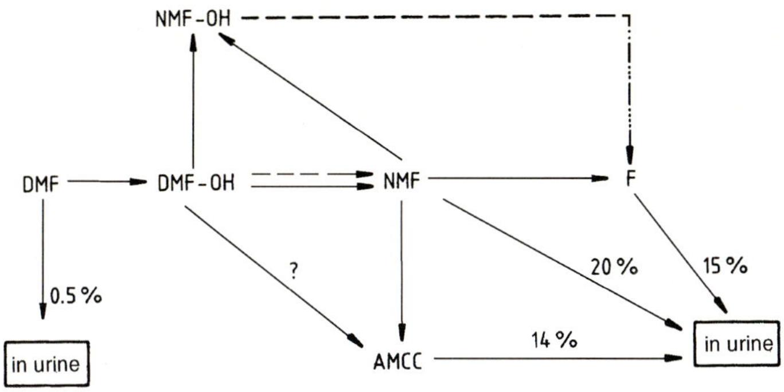

--------------- = GC analytical transformation
———————— = physiological metabolism
DMF = dimethylformamide
AMCC = *N*-acetyl-*S*-(*N*-methylcarbamoyl)-cysteine
DMF-OH = *N*-methyl-*N*-hydroxymethylformamide
F = formamide
NMF = *N*-methylformamide
NMF-OH = *N*-hydroxymethylformamide

Fig. 1: Metabolism of DMF in the human organism taking into account the break-down reaction occurring with GC analysis (modified according to (1))

2 Critical Toxicity

DMF is considered to be a hepatotoxic substance, above all in high dose ranges (8, 14, 20, 21). In doses like those occurring at the workplace increases in the activity of transaminases, LDH and γ-GT as well as an increase in cholesterol, triglycerides and bilirubin have been observed. Even low concentrations of DMF (below 30 mg/m^3) can lead to alcohol intolerance in the sense of a disulfiram-like effect with reddening of the skin of the facial area, thorax and arms, a feeling of warmth, severe headaches, confusion, tightness in the chest, anxiety and dizziness (3, 4, 13).

Further data on toxicity can be found in the documentation of the MAK value (6).

3 Exposure and Effects

Experience with long-term surveillance using biological monitoring of workers exposed to DMF has been published by Lauwerys (12). According to his observations contact with DMF can be regarded as “safe” if in the post shift urine not more than 40–50 mg NMF/g creatinine can be found. Acute or long-term effects on liver function have not been observed below this urine concentration. Alcohol abuse can, however, still lead to a disulfiram-like effect even below 40 mg NMF/g creatinine.

Lyle et al. (13) observed flush reactions after intake of alcohol in persons exposed occupationally to DMF levels below 10 ppm. The DMF metabolite excretion was usually below 10 mg/l. Yonemoto et al. (23) were able to verify these results. Larese et al. (11) reported a positive correlation between the DMF metabolite elimination and changes in the liver function in persons occupationally exposed to DMF in the production of artificial leather. The average NMF concentration amounted to 24 ± 28 mg/l (N = 78) for urine samples collected at the end of the shift. Increased AST, ALT and γ-GT activities occur occasionally with NMF excretion levels of less than 40 mg/l.

4 Selection of the Indicators

The NMF concentration in the post-shift urine sample is regarded as the most suitable parameter for biological monitoring. Care should be taken that during the analytical determination the NMF excreted in the urine as well as the NMF formed from DMF-OH during gas chromatographic analysis is quantified (see sections 1 and 5).

In persons not occupationally exposed to DMF no DMF metabolites are detectable in the urine. Alcohol consumption on the day of the investigation should be avoided as this can cause a decrease in DMF metabolism (22). DMF metabolites can already be detected in urine shortly after uptake. The elimination maximum reached shortly after the end of exposure. The half-life is approx. four hours. The elimination is over within 24 hours (19). Accumulation of the metabolites has not been described. The kinetic processes differ somewhat depending upon the route of absorption; after percutaneous absorption the kinetics are somewhat delayed (18).

The other metabolites and the measurement of renally excreted DMF has gained no importance for kinetic and analytical reasons. However, Mráz and Hohova (18) suggest using AMCC excretion as a biological marker of percutaneous uptake of DMF. The current database for this parameter, however, does not allow evaluation of such a BAT value.

5 Methodology

The quantitative determination of *N*-methylformamide in urine takes place as a rule by gas chromatography. It should be taken into consideration that during the thermal treatment of the sample in the gas chromatograph the hydroxymethyl group of *N*-hydroxymethyl-*N*-methylformamide is cleaved forming *N*-methylformamide which is quantified together with the NMF excreted in small amounts by occupationally exposed persons. Complete thermal cleavage takes place at temperatures of 250 °C in the injector block of the gas chromatograph. This was shown in a study by Kawai et al. (7). At temperatures around 200 °C the NMF levels determined are only a quarter of those found at 250 °C. From 250 °C there is a plateau until 300 °C.

Detailed descriptions of the methods for the determination of *N*-methylformamide in urine can be found by Mráz (17) as well as Kawai et al. (7). Urine is injected directly or diluted before injection and *N*-methylformamide is determined with a flame ionisation detector (FID) (7) or with a nitrogen-specific detector (17). A gas chromatographic method with nitrogen-specific detector and direct measurement has also been put forward by the Working Group "Analysen in biologischem Material" ("Analyses of Hazardous Substances in Biological Materials") of the DFG Commission for the Investigation of Health Hazards of Chemical Compounds in the Work Area (5).

The urine samples are stable for investigation for a long period of time and can be stored for at least a week in the refrigerator without a change in the concentration.

For the determination of AMCC in urine there are methods recommended by Mráz (17) as well as Mráz and Turecek (16). The procedures involve derivatisation of AMCC with ethanol under alkaline conditions. The ethyl-*N*-methylcarbamate thus produced is quantified gas chromatographically.

6 Evaluation of the BAT Value

For the evaluation of the BAT value the study published by Kawai et al. (7) is the main source. The following points speak in favour of this:

1. In this study adequate personal air monitoring was carried out. The authors draw attention to the fact that differences between their data and data from previous studies can be explained up to a point by the fact that exclusively stationary measurement of the air concentration considerably underestimates the exposure.

2. In this study care was taken that skin absorption did not take place. In contrast, in previous studies skin absorption could not be excluded with certainty.

3. The authors were able to prove that the amounts of NMF detectable in the urine increased up to an analysis temperature of 250 °C. Thus, in earlier studies in which the analysis was carried out at 200 °C (9, 10, 15, 23) falsely low concentrations of NMF in urine were measured. In the study published by Lauwerys et al. (12), which can be used as the basis for exposure-effect relationships, gas chromatographic conditions of 250 °C were selected. This study is thus relevant.

A further argument in favour of the study is the size of the collective with 208 exposed test persons.

A linear relationship between DMF in air and NMF in urine was found by Kawai et al. (7). According to the formula derived by the authors, for a MAK value of 10 ppm there is a urinary excretion of 17 mg NMF/l urine or 16 mg NMF/g creatinine.

Krivanek et al. (10) in a study on volunteers also describe a linear relationship. The excretion of 4 mg/d NMF with 10 ppm DMF is, however, greatly below that of Kawai et al. (7), probably due to the method. In workers with corresponding physical activity a two to three-fold higher excretion is expected by the latter authors. From the observations of Yonemoto and Suzuki (23) with exposure to 10 ppm a value of 39 mg NMF total excretion in the 24-hour urine can be calculated. In this investigation with nine workers exposed for eight hours skin absorption could, however, not be excluded. In contrast, Lauwerys et al. (12) deduced on the basis of their investigations a correlation yielding approx. five times higher calculated values of NMF. Kawai et al. (7) account for these results by explaining that possibly the external exposure in the study by Lauwerys was underestimated as exclusively stationary ambient monitoring was carried out. On the basis of the results and the conclusions of Kawai et al. (7) a BAT value of

15 mg *N*-methylformamide/l urine

is suggested for a urine sample collected at the end of the shift.

This value is supported on the one hand by the investigations by Lauwerys et al. (12), who could observe no adverse effects with NMF excretion of <40 mg/g creatinine in persons occupationally exposed to DMF. On the other hand the value is more or less in accordance with the results of various field studies and the results of exposure experiments on volunteers. A table of the results of these studies can be found in the BEI documentation (1). With exposure to a DMF concentration of 10 ppm for eight hours NMF excretion levels of 11–38 mg/d were observed in field studies; in experimental investigations the values varied between 4 and 35 mg/d. Studies in which the urine samples were collected at the end of the shift produced values under field conditions in the range of 18 mg/l to 36 mg/l or 70 mg/g creatinine, under experimental conditions the value was 44 mg/l.

7 Interpretation of the Data

TRGS 150 "immediate skin contact with toxic substances" states that for substances with a systemic effect and available BAT value it should be ensured that the BAT value is observed. *N,N*-Dimethylformamide falls into this category. Biological monitoring should be given preference for occupational medical surveillance due to the good dermal uptake. A berufsgenossenschaftlicher Grundsatz (guideline issued by the Employers' Liability Insurance Association) for *N,N*-dimethylformamide is not available.

In practice it is important that the urine sampling takes place directly after the end of the exposure phase. Due to the stability of the DMF metabolite the urine samples can be stored in the refrigerator for at least a week. For the analysis attention is particularly drawn to the fact that the gas chromatographic determination described in the section "Methodology" should be used. Only then is it guaranteed that indeed the sum of *N*-methylformamide and the *N*-methylformamide formed from *N*-methyl-*N*-hydroxymethylformamide is determined. The established BAT value is based on this method.

In persons not occupationally exposed no DMF metabolites can be detected with the suggested analytical procedure. A non-occupational exposure is improbable. There are or were, however, some household products, such as paint remover and shoe polish, which could have contained DMF.

8 References

1. American Conference of Governmental Industrial Hygenists (ACGIH): Documentation of the Threshold Limit Values and Biological Exposure Indices. BEI 75, 6th ed. ACGIH, Cincinnati, OH 1991
2. Brugnone, F., Perbellini, L., Gaffuri, E.: *N,N*-dimethylformamide concentration in environmental and alveolar air in an artificial leather factory. Br. J. Ind. Med. *37* (1980), 185–188
3. Eben, A., Kimmerle, G.: Metabolism studies of *N,N*-dimethylformamide, III. Studies about the influence of ethanol in persons and laboratory animals. Int. Arch. Occup. Environ. Health *36* (1976), 243–265
4. Hanasono, G. K., Fuller, R. W., Broddle, W. D., Gibson, W. R.: Studies on the effects of *N,N*-dimethylformamide on ethanol disposition and on monoamine oxidase activity in rats. Toxicol. Appl. Pharmacol. *39* (1977), 461–472
5. Henschler, D. (ed.): Analytische Methoden zur Prüfung gesundheitsschädlicher Arbeitstoffe. Band 2: Analysen in biologischem Material, 10th issue. VCH Verlagsgesellschaft, Weinheim 1991
6. Henschler, D. (ed.): Gesundheitsschädliche Arbeitsstoffe., Toxikologisch-arbeitsmedizinische Begründungen von MAK-Werten, 18th issue. VCH Verlagsgesellschaft, Weinheim 1992
7. Kawai, T., Yagusi, T., Mizunuma, K., Watanabe, T., Cai, S., Huang, M., Xi, L., Qu, J., Yao, B., Ikeda, M.: Occupational dimethylformamide exposure, 2. Monomethylformamide excretion in urine after occupational dimethylformamide exposure. Int. Arch. Occup. Environ. Health *63* (1992), 455–460
8. Kennedy, G. L., Jr.: Biological effects of acetamide, formamide, and their monomethyl and dimethyl derivates. CRC Crit. Rev. Toxicol. *17*(2) (1986), 129–182
9. Kimmerle, G., Eben, A.: Metabolism studies of *N,N*-dimethylformamide, II. Studies in persons. Int. Arch. Arbeitsmed. *34* (1975), 127–136
10. Krivanek, N. D., McLaughlin, M., Fayerweathe, W. E.: Monomethylformamide levels in human urine after repetitive exposure to dimethylformamide vapor. J. Occup. Med. *20* (1978), 179–182
11. Larese, F., Fiorito, A., Dobetti, L. et al.: Dimethylformamide liver effects: The role of the skin route. 1st International Scientific Conference on Occupational Hygiene. Brussels, Belgium. Int. Occup. Hyg. Assoc. Dec. (1992) 7–19 (Abstract)
12. Lauwerys, R. R., Kivits, A., Lhoir, M., Rigolet, P., Houbeau, D., Bouchet, J. P., Roels, H. A.: Biological surveillance of workers exposed to dimethylformamide and the influence of skin protection on its percutaneous absorption. Int. Arch. Occup. Environ. Health *45* (1980), 189–203
13. Lyle, W. H., Spence, T. W. M., McKinnley, W. M., Duckers, K.: Dimethylformamide and alcohol intolerance. Br. J. Ind. Med. *36* (1979), 63–66
14. Massmann, W.: Toxicological investigations on dimethylformamide. Br. J. Ind. Med. *13* (1956), 51–54
15. Maxfield, M. E., Barnes, J. E., Azar, A., Trochimowicz, H. T.: Urinary excretion of metabolites following experimental human exposures to DMF or to DMAC. J. Occup. Med. *17* (1975), 506–511
16. Mráz, J., Turecek, F.: Identification of *N*-acetyl-*S*-(*N*-methylcarbamoyl)cysteine, a human metabolite of *N,N*-dimethylformamide and *N*-methylformamide. J. Chromatogr. *414* (1987), 399–404
17. Mráz, J.: Gas chromatographic method for the determination of *N*-acetyl-*S*-(*N*-methylcarbamoyl)cysteine, a metabolite of *N,N*-dimethylformamide and *N*-methylformamide in human urine. J. Chromatogr. *431* (1988), 361–368
18. Mráz, J., Nohová, H.: Percutaneous absorption of *N,N*-dimethylformamide in humans. Int. Arch. Occup. Environ. Health *64* (1992), 79–83
19. Mráz, J., Nohová, H.: Absorption, metabolism and elimination of *N,N*-dimethylformamide in humans. Int. Arch. Occup. Environ. Health *64* (1992), 85–92
20. Scailteur, V., Lauwerys, R. R.: Dimethylformamide (DMF) hepatoxicity. Toxicology *43* (1987), 231–238
21. Tanaka, K. I.: Toxicity of dimethylformamide (DMF) to the young female rat. Int. Arch. Arbeitsmed. *28* (1971), 98–105
22. Tichy, M., Mráz, J.: Solvents interactions and biological monitoring. In: Imbriani/DiNucci (ed.), Effetti della interazione tra etanolo e solventi. La Goliardica Pavese. Pavia, Italien 1991, pp. 67–81
23. Yonemoto, J., Suzuki, S.: Relation of exposure to dimethylformamide vapor and the metabolite, methylformamide, in urine of workers. Int. Arch. Occup. Environ. Health *46* (1980), 159–165

Authors: K. H. Schaller, H. Drexler
Approved by the Working Group: 27th May 1993

Hexachlorobenzene

BAT	150 µg HCB/l plasma
Date of evaluation	1984
Synonyms	Perchlorobenzene
Formula	C_6Cl_6
Molecular weight	284.4
Melting point	230 °C
Boiling point	322 °C

1 Metabolism and Kinetics

During occupational exposure HCB absorption is predominantly pulmonary and percutaneous. Due to its physico-chemical properties HCB is accumulated in tissues containing fat. Investigations on rhesus monkeys and pigs revealed increased HCB concentrations in the body fat, bone marrow, kidneys, brain and thymus (1–3). In the liver of rats the HCB concentration was 10–20 times that in plasma (4). A ratio of 1:500 was found for the HCB concentration in plasma to that in adipose tissue (3, 4). Adipose tissue and liver represent the largest compartments of the HCB pool.

The biotransformation of HCB takes place mainly in the liver with the help of the microsomal enzymes. In rats after exposure to HCB the activity of the hydroxylating enzyme cytochrome P 450 in the liver increased to three times its normal activity (5).

In rhesus monkeys and rats pentachlorophenol (PCP) was identified as the main metabolite of HCB (2, 6–8). According to investigations with ^{14}C-HCB, PCP made up approx. 20 % of the metabolites excreted in the urine (9). In addition tetrachlorohydroquinone, pentachlorothiophenol, pentachlorothioanisole, 2,4,5-trichlorophenol, 1,2,4,5-tetrachlorobenzene, 2,3,4,6-tetrachlorophenol and other unidentified chlorophenols and chlorobenzenes occur as metabolites.

From the results of various investigations of human adipose tissue (10, 11) it can be concluded that a steady state between the HCB uptake and the HCB elimination develops at a level proportional to the HCB exposure and the size of the HCB pool (12). The mechanisms producing the steady state prevent a continuous HCB accumulation in the body (10).

In what way the human organism eliminates absorbed HCB in unchanged form has so far not been investigated. From the results of animal experimental studies it can, however, be concluded that the excretion of metabolised HCB takes place mainly via the walls of the intestinal tract (9). HCB excretion in urine, on the other hand, is negligible.

There is little data on the rate of elimination of absorbed HCB, on the biological half-life, and the results which are available are almost exclusively from animal experiments and are very variable. For rats after long-term exposure to HCB under steady-state conditions a half-life of 4–5 months was given (6). With values for the young steers of 10.5 weeks (13) and for laying hens of 24–27 days (14), the relatively short duration of the HCB administration (10 weeks and 7 days, respectively), seems insufficient to ensure that the determination of the biological half-life occurred from stable steady-state conditions. Such conditions were only reached with comparable doses in rats after 9 weeks (6).

In workers exposed to HCB, the biological half-life was found to be approx. 2 years (15).

2 Critical Toxicity

HCB is of low acute toxicity. For rats given HCB by oral application LD_{50} values of 3.5–10 g/kg body weight have been published (16).

In rats the administration of HCB led to increased formation and excretion of uroporphyrinogen and coproporphyrin as well as – after longer exposure – of porphobilinogen (3–6, 17–19). Increased excretion of delta-aminolaevulinic acid (δ-ALA) has also been detected (20).

The porphyria occurring in animal experiments after HCB application could only be confirmed to a certain extent in persons exposed to HCB. Investigations on 54 Australian workers who treated seed with HCB and distributed it resulted in moderately increased excretion of uroporphyrin (170 µg/d) and coproporphyrin (255 µg/d) only in a 64-year-old after 25 years of exposure. Measurements of the HCB blood levels are not available for these persons (21).

In 20 persons of a spraying troupe, some of whom had used a herbicide contaminated with HCB for many years, HCB determination in plasma revealed on average 0.04 ppm (range 0–0.31 ppm). The HCB load had, however, no effect on biological functions or biochemical parameters; in particular there were no effects on the uroporphyrin and coproporphyrin excretion. Merely the δ-ALA excretion in the urine was correlated with the HCB blood level which, however, was evaluated with reservations (22).

A group of 50 chlorinated hydrocarbon production workers were monitored from 1974–1977. The average HCB plasma level fluctuated during the observations around 0.325 ppm (15). This level corresponds to 65 times the average blood HCB concentration observed in the general population (see review of the published HCB concentrations in whole blood and plasma from occupationally exposed and non-occupationally exposed persons in Tab. 1 and 2). Despite these high internal levels of HCB, biological parameters, in particular those of porphyrin metabolism and liver function, lay within the normal range and showed no statistically significant differences from the data for a control collective without occupational HCB exposure (15).

In farmers who had used pesticides containing HCB over a long period, HCB levels in the blood of 40 ppb were measured. They showed no recognisable health effects (22).

In 35 persons exposed to pesticides, in comparison to persons not exposed, 20 times higher HCB blood values were measured than in unexposed persons (23).

Clinical experience of HCB intoxication has been reported only from an epidemic which occurred during the years 1955–1959 in south eastern Turkey and which affected an estimated 4000 persons (24). The substance was taken up with the food in amounts of 0.05–0.2 g HCB/person/day during several months (16, 24–26). Approx. 4000 cases of porphyria cutanea tarda were registered. The acute clinical symptoms regressed within 20–30 days after the HCB intake was recognised as the cause and stopped. A follow-up investigation of those affected after 20 (27) and 25 years (28) revealed abnormal porphyrin metabolism in 32 persons. The mechanism by which HCB induces porphyria is still unclear. As none of the main HCB metabolites can induce pathological porphyria reactive intermediates are probably responsible.

3 Exposure and Effects

Exposure

According to the few data available so far for persons occupationally exposed to HCB the HCB concentrations in whole blood or plasma must be regarded as an exposure parameter.

Effect

An abnormal porphyrin metabolism with increased porphyrin excretion is to be regarded as a parameter of effect with HCB exposure. A quantitative relationship between exposure and effect could, however, so far not be demonstrated.

4 Selection of the Indicators

a) Porphyrin Excretion

In animal experiments, after HCB exposure increased uroporphyrin and coproporphyrin excretion was measured. This could be confirmed only to a certain extent in persons exposed to HCB. In the investigations available, values for coproporphyrin between 50–300 µg/l urine and for uroporphyrin between 10 and 40 µg/l urine were given (15, 21). Only in individual cases was the porphyrin excretion not in the normal range.

It must be taken into consideration that increased porphyrin excretion is in no way specific for HCB exposure.

b) HCB in Blood

In persons exposed to HCB considerably increased HCB values in blood were measured compared to those in non-exposed control persons (c.f. section 2). The measurement of HCB in blood is therefore a specific parameter for HCB exposure. During monitoring over several years of persons exposed to HCB a good correlation between the HCB concentration and the duration of exposure was observed. The correlation between the blood HCB concentration and the air concentration is on the other hand poor (15).

5 Methodology

HCB in Blood

A gas chromatographic method is described for the determination of HCB in blood (23). A summary of the methods available for HCB is given by Li et al. (1976) (29).

The working group “Analyses of Hazardous Substances in Biological Materials” have developed a reliable method and published it in their loose-leaf collection of methods (30).

6 Background Exposure

Table 1 contains data on HCB concentrations in blood and plasma samples of persons from the general population. On the basis of modern methods of analysis the upper limit of the normal range is given as 4 µg/l for plasma (31) and 1.2 µg/l for whole blood (32).

Table 2 provides an overview of HCB concentrations in blood in persons with occupational HCB exposure.

7 Evaluation of the BAT Value

In the Federal Republic of Germany and in other industrial states MAK/TLV values for HCB are not available. Recently the American Conference of Governmental Industrial Hygienists have issued a notice of intent to change to a TLV value of 0.025 mg/m^3. Only in the Hygiene Standard List of the former USSR is HCB included. At 0.9 mg/m^3 HCB is high compared to that for other substances in the list.

The “FAO-WHO Food Standards Programme Codex Committee on Pesticide Residues” has set an ADI (allowable daily intake) value of 0.6 µg HCB/kg body weight/day (= 45 µg/day for a 75 kg person). This amount, considered by the WHO to be without cause for concern, is ten times higher than the currently normal HCB uptake (approx. 4.5 µg/day) (33). The uptake of 4.5 µg HCB/day leads to a HCB-blood value of 4 µg/l plasma, the uptake of 45 µg HCB/day corresponds to 40 µg/l plasma.

In the light of the occupational medical investigations in which HCB values up to 310 µg/l were measured without recognisable health effects a BAT value for HCB, including a safety factor, is set at

150 µg HCB/l plasma.

8 Interpretation of the Data

The BAT value is conceived as an individual maximum value and in the individual case should not at any time - and that is during the whole of a working life – be exceeded. In the evaluation of the results of plasma analyses therefore the tendency to accumulate and long half-life (2 years and longer) of HCB must be taken into consideration.

There is no guideline issued by the Employers' Liability Insurance Association.

9 References

1. Mueller, W. F., Iatropoulos, M. J., Rozman, K., Korte, F., Coulston, F.: Toxicol. Appl. Pharmacol. *45* (1978), 283–284
2. Yang, R. S. H., Pittman, K. A., Rourke, D. R., Stein, V. B.: J. Agric. Food Chem. *26* (1978), 1076–1083
3. Tonkelaar, E. M. den, Verschuuren, H. G., Bankovska, DeVries, T., Kroes, R., Esch, G. J. van: Toxicol. Appl. Pharmacol. *43* (1978), 137–145
4. Kuiper-Goodman, T., Grant, D. L., Moodie, C. A., Korsrud, G. O., Munro, I. C.: Toxicol. Appl. Pharmacol. *40* (1977), 529–549
5. Lissner, R., Koerz, G., Eichenauer, M. G., Ippen, H.: Biochem. Pharmacol. *24* (1975), 1727
6. Koss, G., Seubert, S., Koransky, W., Ippen, H.: Arch. Toxicol. *40* (1978), 285–294
7. Engst, R., Macholz, R. M., Kujawa, M.: Bull. Environ. Contam. Toxicol. *16* (1976), 248–252
8. Schuster, K. P., Rener, G.: N. Arch. Exp. Pathol. Pharmacol. *297* (1977), R 5
9. Koss, G., Koransky, W., Steinbach, K.: Arch. Toxicol. *35* (1976), 107–114
10. Abbott, D. C., Collins, G. B., Goulding, R., Hoodless, R. A.: Brit. Med. J. *283* (1981), 1425–1428
11. Richter, E., Schmid, A.: Arch. Toxicol. *35* (1976), 141–147
12. Grant, D. L., Iverson, F., Hatina, G. V., Villeneuve, D. C.: Toxicol. Appl. Pharmacol. *29* (1974), 101
13. Deichmann, E. B. in: Patty F. A. (ed.): Industrial Hygiene and Toxicology, Vol. 28. Wiley & Sons, New York 1981
14. Hansen, L. G., Dorn, S. B., Sundlof, S. M., Vogel, R. S.: J. Agric. Food Chem. *16* (1978), 1369
15. Currier, M. F., McClimans, C. D., Barna-Lloyd, G.: Toxicol. Environ. Health *6* (1980), 367–377
16. International Agency for Research on Cancer (IARC): The evaluation of the carcinogenic risk of chemicals to humans. IARC Monographs No. 20. IARC, Lyon 1979, pp. 155–178
17. Schmid, R.: New Engl. J. Med. *263* (1960), 397–398
18. Ockner, R. K., Schmidt, R.: Nature *189* (1961), 499
19. Goerz, G., Viethum, W., Bolsen, K., Krieg, Th., Lissner, R.: Arch. Dermatol. Res. *263* (1978), 189–196
20. Ariyoshie, T., Idegichi, K., Ishizika, Y., Iwasaki, K., Akakan, M.: Chem. Pharm. Bull. *23* (1975), 817
21. Morley, A., Geary, D., Harben, F.: Med. J. Aust. *3* (1973), 565
22. Burns, J. E., Miller, F. M., Gomes, E. D., Albert, R. A.: Arch. Environ. Health *29* (1974), 192–194

23. Lunde, G., Bjorseth, A.: Sci. Total Environ. *8* (1977), 241–246
24. Cam, C., Nigogosyan, G.: J. Am. Med. Assoc. *183* (1963), 88–91
25. Cetingil, A. L., Ozen, M. A.: Blood *16* (1960), 1002
26. De Matteis, F., Prior, B. E., Rimington, C.: Nature *191* (1961), 363
27. Cripps, D. J., Peters, H. A., Cocmen, A.: J. Invest. Dermatol. *71* (1978), 277
28. Peters, H. A., Cocmen, A., Cripps, J., Bryan, G. T., Dogramaci, I.: Arch. Neurol. *39* (1982), 744–749
29. Li, R. T., Spigarelli, J. L., Going, J. E.: Sampling and Analysis of Selected Toxic Substances. Task 1 A – Hexachlorobenzene, EPA-560/6/76/001 (1976). US Environmental Protection Agency, Office of Toxic Substances, Report No. PB-253974, available from Springfield, VA National Technical Information Service
30. Angerer, J., Schaller, K. H.: Analyses of Hazardous Substances in Biological Materials, Vol. 3. Deutsche Forschungsgemeinschaft, VCH Verlagsgesellschaft, Weinheim 1991
31. Schaller, K.H., Angerer, J., Lehnert, G.: Bio-Monitoring in der Arbeits- und Umweltmedizin. Deutsches Ärzteblatt – Ärztliche Mitteilungen, 90. Jahrgang, *31/32* (1993), 2122–2128
32. Eckrich, W., Gerhard, I.: Organochlorverbindungen im Blut der Bevölkerung – ein Überblick (Organochlorine Compounds in the Blood of the Population: A Review). Klin. Lab. *38* (1992), 462–468
33. Zimmerli, B., Marek, B.: Lebensmitteluntersuchungen u. Hygiene. Eidg. Drucksachen- und Materialzentrale, Bern, Suisse, 1973
34. Siyali, D. S.: Med. J. Aust. *2* (1972), 1063–1066
35. Siyali, D. S., Ouw, K. H.: Med. J. Aust. *3* (1973), 908–909
36. Burns, E., Miller, F. M.: Arch. Environ. Health *30* (1975), 44–48
37. Eckenhausen, F. W., Bennet, D., Beynon, K. I., Elgar, K. E.: Arch. Environ. Health *36* (1981), 81–92

Tab. 1: HCB concentrations in blood of adults (WB = whole blood, S = serum, P = plasma)

Land	n	Matrix	$\bar{x}$ (ppb)	± s (ppb) (range)	Author
AUS (town)	52	WB	22	(0–95)	Siyali, 1972 (34)
AUS (country)	47	WB	59	(1–179)	Siyali et al., 1973 (35)
CH	pool	S	4.0		Zimmerli et al., 1973 (33)
USA	43	P	0.5	(0–1.8)	Burns et al., 1975 (36)
USA*	86	P	3.57	± 4.3 (0–23)	Burns et al., 1975 (36)
NL	48 (f)	WB	4.9 (ante partum)	(1.9–40)	Eckenhausen et al., 1981 (37)
	73 (f)	WB	5.8 (post partum)		
D	1851	WB	0.062	(0.003–10.4)	Eckrich and Gerhard, 1992 (32)

* Persons living in the vicinity of a road used for transport of chemical waste containing HCB

Tab. 2: HCB concentrations in blood of groups of persons occupationally exposed to HCB (WB = whole blood, P = plasma)

Land	n	Matrix	$\bar{x}$ (ppb)	± s (ppb) (range)	Authors
AUS (town)	185	WB	55.5	(0–410)	Siyali, 1972 (34)
AUS (country)	29	WB	57.1	(5–168)	Siyali et al., 1973 (35)
USA	20	P	40	(1–310)	Burns et al., 1974 (22)
USA	50	P	310.7	± 287.7	Currier et al., 1980 (15)
	49	P	311.5	± 242.9	
	49	WB	159.9	± 142.7	
	44	WB	170.3	± 111.8	

Authors: E. Schumacher-Wittkopf, G. Lehnert
Approved by the Working Group: 17. 2. 1984

γ-1,2,3,4,5,6-Hexachlorocyclohexane

BAT	20 µg γ-1,2,3,4,5,6-hexachlorocyclohexane/l blood 25 µg γ-1,2,3,4,5,6-hexachlorocyclohexane/l serum Sampling time: end of shift after several working days
Date of evaluation	1988
Synonyms	Lindane γ-HCH Gammexane® Benzene-γ-hexachloride
Formula	$C_6H_6Cl_6$

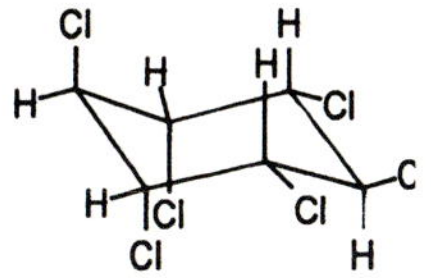

Molecular weight	290.83
Melting point	112.5–113 °C
Boiling point	323.4 °C (1013 hPa)
Vapour pressure at 20 °C	$12 \cdot 10^{-6}$ hPa
MAK [last established 1958]	0.5 mg/m^3

1 Metabolism and Kinetics

1.1 Distribution

Lindane is a lipid-soluble substance with a distribution coefficient (hexane/water) of 1738 (27). It accumulates particularly in adipose tissue.

In the general population lindane is found in the body fat. Details can be found in the literature (6, 18, 19, 20, 36, 40). The relationship between the concentrations in adipose tissue and in blood is given as 30:1 (6). From investigations on the lindane distribution in the blood at concentrations of 10 and 59 µg/l it was discovered that the distribution ratio for lindane between plasma and erythrocytes is 2:1 (28). If the average plasma volume is set at 55 % of the total blood volume (26) then the conversion factor for blood to plasma is 1.22 and for plasma to blood 0.82.

1.2 Metabolism

The metabolism diagram shown in figure 1 was developed from animal experimental studies (39).

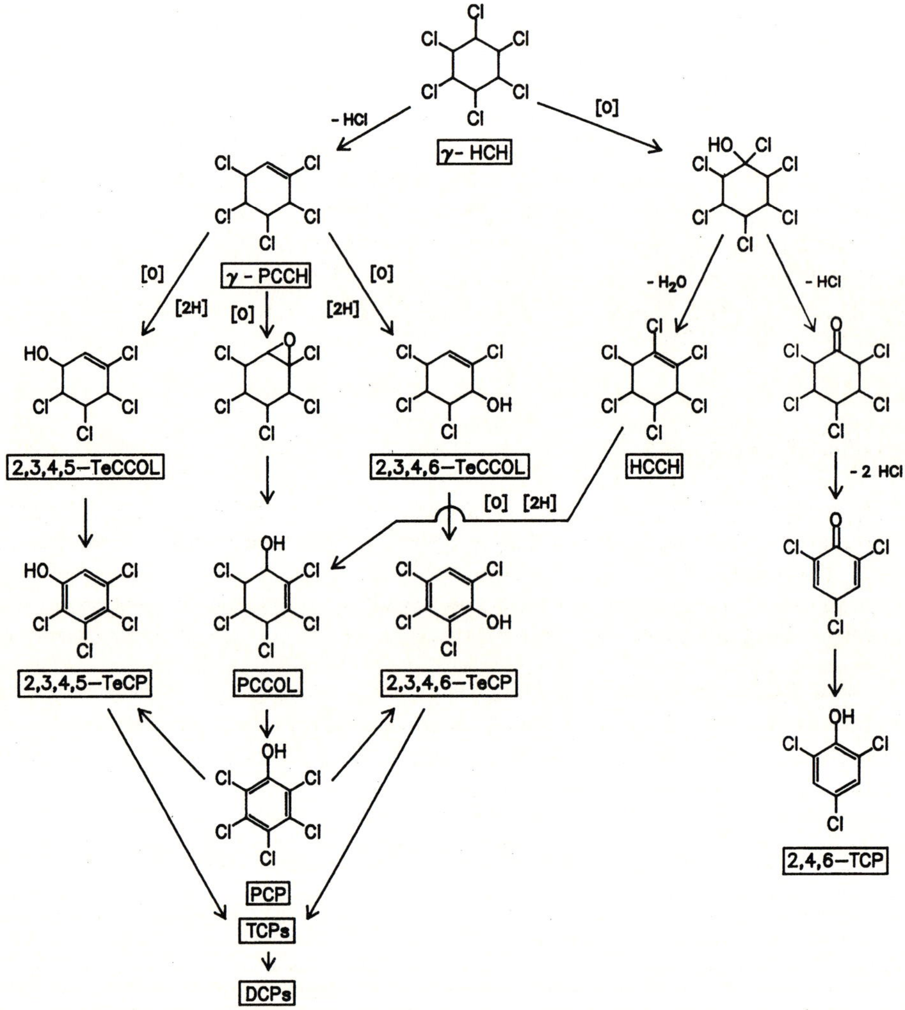

Fig. 1: Metabolism of lindane in mammals (from (39)). Abbreviations: γ-PCCH, γ-2,3,4,5,6-pentachloro-1-cyclohexene; HCCH, hexachlorocyclohexene; PCCOL, 2,3,4,5,6-pentachloro-2-cyclohexene-1-ol; PCP, pentachlorophenol; TeCCOL, 2,3,4,5-tetrachloro-2-cyclohexene-1-ol; TeCP, 2,3,4,5-tetrachlorophenol; TCPs, trichlorophenol isomers; DCPs, dichlorophenol isomers

Angerer et al. (3) investigated the metabolism of hexachlorocyclohexane with exposed employees. The external exposure was mostly to lindane; the amount of α-HCH and β-HCH was small. In the urine there were conjugates of monochlorophenols, dichlorophenols, trichlorophenols and tetrachlorophenols. The main metabolites were the conjugates of 2,4-dichlorophenol, 2,4,6-trichlorophenol, 2,3,5-trichlorophenol and 2,4,5-trichlorophenol.

1.3 Half-lives, Toxicokinetics

Lauwerys (1983) gives a half-life for lindane in blood of 20 hours after short-term exposure (25). Likewise after cutaneous application in children a half-life in the blood of 21 hours was found (13). On the basis of these biological half-lives an estimation of the time needed to reach the steady-state levels can be made. These relationships lead to the accumulation of lindane during the working week. A different situation is revealed in studies which describe chronic exposure conditions. Angerer et al. (2) determined in employees of a factory producing lindane an average half-life of approx. 10 days. This corresponds in its order of magnitude with the observations made by Drummond et al. (10) who describe an average half-life for the lindane concentration of approx. 8 days. It can be concluded that lindane is distributed over several compartments in the organism.

2 Critical Toxicity

In an evaluation by the Deutsche Forschungsgemeinschaft (15) acute uptake of 10–20 mg lindane/kg is described as dangerous to life. Other published lethal doses (14) lie considerably above this value.

Symptoms of lindane intoxication include headache, muscle pains, nausea, vomiting, fatigue, high blood pressure, tremor, ataxia, convulsions, increased respiratory activity (14, 22); and with severe intoxication also tonoclonic spasms (22).

In employees chronically exposed to lindane, changes in clinical-chemical parameters (creatinine, reticulocytes, leucocytes) were reported (29).

The critical target organ of the toxic effect of lindane for the evaluation of a threshold limit value is the central nervous system.

Lindane has stimulating effects on the central nervous system (21). It produces in rats after intraperitoneal application of 45 mg/kg (approx. 50 % of the lethal dose) generalised clonic spasms (31). This effect was also observed after an oral dose of 60 mg/kg (31). In animals with generalised spasms 6–7 µg lindane per g brain were found (31). Tusell et al. (35) found in investigations on rats that after an average effective dose (oral) of 84 mg/kg the concentration of lindane in blood was 1.4 µg/ml, in the brain 5.0 µg/g. The range for the triggering of tonic spasms is given as 1.4–1.6 µg/ml in blood and as 4.9–5.3 µg/g in the brain.

According to the results of animal experiments, after the application of HCH isomers a condition develops in which the individual convulsants have a reduced convulsive effect (16). This effect seems partly to result from accelerated biotransformation of the convulsant, partly however from a decrease in the ability of the central nervous system to be stimulated by HCH isomers (30). Vohland et al. (38) investigated in rats the effect of lindane on spasms induced by pentylenetetrazole (26). The minimum effective dose required to antagonise the effect of pentylenetetrazole, that is, to delay the triggering of clonic spasms, was

4–5 µg lindane per g brain (wet weight). It can be concluded that lindane causes membrane changes in excitatory axons even below the threshold dose for inducing spasms.

Given this neurophysiological situation, recalculation for the relative weights in humans yields an anticonvulsive dose of 6.5 mg total lindane in the brain. The distribution ratio between whole blood and the brain is given as 1:5 (average, calculated from data in (31)) and 1:4 (35) for the rat and as 1:4 for man (37).

EEG changes which are caused by lindane are discussed in section 6.

3 Exposure and Effects

In the Federal Republic of Germany the use of technical grade HCH is banned, only the use of pure γ-HCH is allowed. Therefore it can be assumed in the Federal Republic of Germany that employees are not exposed to toxicologically relevant doses of the isomers of γ-HCH. The use of technical grade hexachlorocyclohexane leads not only to exposure to lindane but also to the other isomers, in particular α-HCH and β-HCH. The toxicokinetic behaviour of these admixtures differ from those of lindane. The large number of chemicals which may be used in the production of lindane, as well as the different toxic behaviour of the admixtures, make the evaluation of exposure and the effect on employees exposed to lindane more difficult.

There are two studies of persons exposed to HCH which investigated the relationship between lindane exposure levels and the occurrence of its metabolites (3, 10). It was observed that there is a significant correlation ($r = 0.74$) between the concentration of 2,3,4,6-tetrachlorophenol in urine and the lindane concentration in serum (3). The ratio between lindane in serum and 2,3,4,6-tetrachlorophenol in urine is given as 1:6.

In the study by Drummond et al. (10) of employees exposed to lindane it was discovered that for the individual trichlorophenol (TCP) metabolites (2,3,5-TCP, 2,4,5-TCP, 2,4,6-TCP) there is no statistically significant correlation between the TCP concentration in the urine and the lindane concentration in plasma. For the sum of the trichlorophenol metabolites a correlation could, however, be found ($r = 0.46$) between the urinary concentration and the lindane concentration in the plasma.

Indicators of the effect on the target organ (CNS) of the critical toxicity are difficult to evaluate under practical conditions of exposure. In individual studies physiological parameters of brain metabolism and the transmission of impulses are discussed. Electroencephalography is at present the only method of determination of possible changes in cortical activity in persons exposed to lindane. This technique has also been used within the framework of studies on persons exposed to lindane. Further reference is made to this in section 6.

4 Selection of the Indicators

In principle both the lindane concentration in the blood or serum/plasma and the metabolite excretion (chlorophenols) are available as parameters of exposure (3, 4, 5, 9, 10, 23, 29, 32). Recent studies have shown that the lindane concentrations in blood or serum samples from the general population are below 0.1 µg/l (11, 33).

Sufficient published data for the evaluation of a BAT value are, however, only available for the parameter lindane in blood (whole blood/ or serum/plasma). Parameters of effect which could be put to practical use are at present not available (see section 3).

5 Methodology

The lindane concentration in the serum can be analysed by gas chromatography. A tested and reliable method of determination has been published in the collection "Analysen in biologischem Material" ("Analyses of Hazardous Substances in Biological Materials") (1).

6 Evaluation of the BAT Value

Quantifiable relationships between the external occupational exposure and indicators of an internal exposure or effect belong to the criteria for the evaluation of a BAT value. The published investigations for lindane only fulfil these prerequisites to a limited extent as often the exposure was to a mixture of HCH isomers or such admixtures could not be excluded. On the other hand at the time of the most intensive use of lindane or technical grade HCH in the industrialised states (1950s and 1960s) investigations on disturbances of the central nervous system in relation to the external and internal lindane exposure were rare. As already mentioned in section 3 studies on the EEG changes under field conditions are available. The discussion of the findings is difficult for the following reasons: disturbances which lead to changes in the amplitude, frequency or reactivity can have many causes. EEG changes are thus to be regarded as not specific for the effects of a toxic substance. There are non-specific EEG patterns which cannot be attributed to a specific pathophysiological reaction or a specific disease process (34). They represent a reaction to systemic, toxic or metabolic disturbances but can also occur as symptoms of diffuse or focal intracranial diseases. Non-specific dysrhythmic changes were also found in persons ("normal" population) who show no clinical anomalies and are regarded as healthy (12). Therefore it is particularly important that EEG examinations are carried out before the beginning of exposure in order to be able to determine whether or not a dysrhythmia of this kind already existed at this time. Only in the investigations by Czeglédi-Jankó and Avar (9) was a preliminary examination of this kind carried out in one person of a total 15 persons exposed to lindane (without a positive result); in other investigations this preliminary examination did not take place.

After acute and chronic intoxication due to chlorinated hydrocarbon insecticides a deceleration of the basis rhythm to 6 Hz was observed (24).

Paroxysmal bilateral synchronous theta activity was found in 10 % of the employees (n = 122) of a factory producing insecticide (no hexachlorocyclohexane) (17). There was no information on the level of exposure to aldrin, dieldrin, endrin or isodrin. Epileptic-like changes in the sense of a bilateral synchronous spikes and wave complexes were observed in 6 % of the employees in this study (17).

In order to be able to establish a threshold limit value for the internal lindane exposure it is necessary to tabulate parameters of internal exposure (lindane-blood/serum/plasma concentration) and the EEG changes observed.

Czeglédi-Jankó and Avar (9) investigated employees exposed to lindane for between half a year and 2 years in a factory producing fertilisers. No information was given on the purity of the lindane or the possible admixtures, and the level of the external exposure. Of a total of 37 persons investigated 15 were without previous exposure to aldrin. The average duration of exposure of this collective was 1.3 years, the blood lindane concentration varied from 4 to 160 µg/l. The arithmetic mean was 38 µg/l, the geometric mean 19 µg/l.

EEG changes occurred at lindane concentrations in blood of 16 µg/l or more. The percentage of persons with blood lindane levels greater than or equal to 16 µg/l blood was 53 % of the collective. Mild clinical symptoms were described in persons with blood levels of 24 µg/l or more (9).

Drummond et al. (10) observed in 4 % of 45 employees who handled lindane in the agricultural sector non-specific symptoms (flu-like indisposition, tiredness, inflammation of the throat, nausea) which were regarded as symptoms of intoxication. The lindane plasma concentrations were 22 and 36 µg/l. After the end of this exposure the symptoms persisted for approx. 14 days. At the time when the employees first became free of symptoms plasma lindane concentrations of 7 and 8 µg/l were measured. Baumann et al. (5) carried out EEG examinations with 8 of 60 employees of a factory producing lindane. Three persons were not taken into consideration in the evaluation as in

Tab. 1: Values of the parameters for sensitivity, specificity and validity

	Sensitivity	Specificity	Validity	Study
Clinical symptoms	0.71	0.85	1.56	Z
	0.43	1.00	1.43	L
EEG changes	0.88	0.95	1.83	Z
	1.00	0.88	1.88	L
	0.20	0.33	0.55	B
EEG changes (corrected*)	0.88	0.85	1.73	Z
	1.00	0.88	1.88	L
Clinical symptoms or EEG changes (not corrected) or both	1.00	0.80	1.80	Z
	1.00	0.88	1.88	L

Z = Total collective of the study by Czeglédi-Jankó and Avar (9) calculated by Zielhuis and Verberk (41), L = collective exposed exclusively to lindane in the study by Czeglédi-Jankó and Avar (10), B = Baumann et al. (5); *) Zielhuis and Verberk (41) carried out a correction in which they took into account the number of cases to be expected on the basis of abnormal non-specific EEG changes in persons not occupationally exposed. A correction of the part of the collective exposed to lindane whose upper exposure limit is the normal limit for non-occupationally exposed persons was not necessary as the number of cases expected coincided with the number of observed cases (it was assumed that 16 % of persons not exposed occupationally show non-specific EEG changes (12)).

their case previous damage was determined anamnestically. In the other 5 persons a normal α-rhythm was observed, in one person under conditions of hyperventilation a theta-rhythm. The lindane concentration in the serum of this worker was 23 μg/l. The lindane concentrations in the serum of the 5 persons investigated were 10–71 μg/l. Baumann et al. (4) also carried out electrophysiological investigations of the peripheral nervous system in this collective. They found no statistically significant differences from the control persons. The lindane concentrations in the serum varied from 5 to 188 μg/l with a median value of 23 μg/l.

Zielhuis and Verberk (41) investigated the validity parameters, sensitivity and specificity for the indicators "clinical symptoms", "EEG changes" and "clinical symptoms or EEG changes or both" (Tab. 1) for the whole collective of the study by Czeglédi-Jankó and Avar (9). For the collective which had contact only with lindane, the values from the study (41) are shown in table 1. In table 1 the results of the study by Baumann et al. (5) are also included.

On the basis of the validity values for the individual indicators Zielhuis and Verberk (41) regard the combination of the clinical symptoms and an evaluation of the EEG as a criteria for the effects of lindane. Evaluation of the study by Baumann et al. (5) shows no correlation between EEG changes and hexachlorocyclohexane (α, β, γ-HCH) exposure.

On the basis of these results a BAT value of

20 μg γ-1,2,3,4,5,6-hexachlorocyclohexane/l blood
or
25 μg γ-1,2,3,4,5,6-hexachlorocyclohexane/l serum

has been set.

The blood samples should be taken at the end of the shift after several working days.

7 Interpretation of the Data

Due to the ban on the use of technical grade HCH and the limited ban on the use of lindane as a substance for treating plants (22) as well as the currently valid regulations for the maximum levels of pesticides in foodstuffs and tobacco products (7, 8) at present in the Federal Republic of Germany only a slight exposure of the general population can be demonstrated. Schaller et al. (33) discovered for the investigation period 1986 in 114 persons that the plasma lindane concentration in 74 % of those investigated from the general population was below the detection limit of 1 μg lindane/l plasma (see also section 5).

8 References

1. Angerer, J. in: Henschler, D. (ed.): Analytische Methoden zur Prüfung gesundheitsschädlicher Arbeitsstoffe. Bd. 2: Analysen in biologischem Material. 7th issue. VCH Verlagsgesellschaft, Weinheim 1983
2. Angerer, J., Behling, K., Brassow, H.-L., Lehnert, G. in: Bericht über die 23. Jahrestagung der Deutschen Gesellschaft für Arbeitsmedizin. Gentner, Stuttgart 1983, pp. 445–450
3. Angerer, J., Maaß, R., Heinrich, R.: Int. Arch. Occup. Environ. Health *52* (1983), 59–67
4. Baumann, K., Angerer, J., Heinrich, R., Lehnert, G.: Int. Arch. Occup. Environ. Health *47* (1980), 119–127
5. Baumann, K., Behling, K., Brassow, H.-L., Stapel, K.: Int. Arch. Occup. Environ. Health *48* (1981), 165–172
6. Bertram, H. P., Kemper, F. H., Zenzen, Ch. in: Deutsche Forschungsgemeinschaft (ed.): Hexachlorcyclohexan als Schadstoff in Lebensmitteln. Forschungsbericht. VCH Verlagsgesellschaft, Weinheim 1983, pp. 155–163
7. Bundesminister für Jugend, Familie und Gesundheit: Bundesgesetzblatt Teil 1, 1982, pp. 745–784
8. Bundesminister für Jugend, Familie und Gesundheit: Bundesgesetzblatt Teil 1, 1984, pp. 635–643
9. Czeglédi-Jankó, G., Avar, P.: Br. J. Ind. Med. *27* (1970), 283–286
10. Drummond, L., Gillanders, E. M., Wilson, H. K.: Br. J. Ind. Med. *45* (1988), 493–497
11. Eckrich, W., Gerhard, I.: Organochlorverbindungen im Blut der Bevölkerung – ein Überblick. Klin. Lab. *38* (1992), 462–468
12. Gibbs, F. A., Gibbs, F. L., Lennox, W. G.: Arch. Neurol. Psychiat. *50* (1943), 111–128
13. Ginsburg, C. M., Lowry, W., Reisch, J. S.: J. Pediatr. *91* (1977), 998–1000
14. Gosselin, R. E., Smith, R. P., Hodge, H. C., Braddock, J. E.: Clinical Toxicology of Commercial Products. 5th ed. Williams & Wilkins, Baltimore 1984, pp. III-239 – III-241
15. Herbst, M. in: Deutsche Forschungsgemeinschaft (ed.): Hexachlorcyclohexan als Schadstoff in Lebensmitteln. Forschungsbericht. VCH Verlagsgesellschaft, Weinheim 1983, pp. 191–200
16. Herken, H.: Naunyn-Schmiedeberg's Arch. Exp. Path. Pharmakol. *221* (1956), 143–150
17. Hoogendam, I., Versteeg, J. P. J., de Vlieger, M.: Arch. Environ. Health *4* (1962), 92–100
18. International Agency for Research on Cancer (IARC): The Evaluation of the Carcinogenic Risk of Chemicals to Humans. IARC Monographs No. 5: Some Organochlorine Pesticides. IARC, Lyon 1973, pp. 47–74
19. International Agency for Research on Cancer (IARC): The Evaluation of the Carcinogenic Risk of Chemicals to Humans. IARC Monographs No. 20: Some Halogenated Hydrocarbons. IARC, Lyon 1979, pp. 195–239
20. Jani, J. P., Patel, J. S., Shah, M. P., Variya, M. R., Shah, Y. H., Gupta, S., Kashyap, S.: Scand. J. Work Environ. Health *14* (1988), 201–204
21. Joy, R. M.: Neurobehav. Toxicol. Teratol. *4* (1982), 813–823
22. Klaassen, C. D., Amdur, M. O., Doull, J.: Casarett and Doull's Toxicology, 3rd ed. Macmillan, New York 1986, pp. 549–551
23. Kolmodin-Hedman, B., Palmér, L., Götell, P., Skerfoing, S.: Work Environ. Health *10* (1973), 100–106
24. Kurtz, D. in: Remond, A. (ed.): Handbook of Electroencephalography and Clinical Neurophysiology. Vol. 15 C. Elsevier, Amsterdam 1976, pp. 105–107
25. Lauwerys, R. R.: Industrial Chemical Exposure: Guidelines for Biological Monitoring. Davis: Biomedical Publ. (1983), 116
26. Lentner, C.: Geigy Scientific Tables. Vol. 3: Physical Chemistry, Composition of Blood, Hematology, Somatometric Data. 8th ed. Ciba-Geigy, Basel 1984, p. 65
27. Leo, A., Hansch, C., Elkins, D.: Chem. Reviews *71* (1971), 525–616
28. Lewalter, J.: Personal communication, 1987
29. Milby, T. H., Samuels, A. J.: J. Occup, Med. *13* (1971), 256–258
30. Portig, J., Koransky, W., Walmer-Roedler, D.: Naunyn-Schmiedeberg's Arch. Exp. Path. Pharmcol. *274* (1972), 154–170
31. Portig, J., Vohland, H. W. in: Deutsche Forschungsgemeinschaft (ed.). Hexachlorcyclohexan als Schadstoff in Lebensmitteln. Forschungsbericht. VCH Verlagsgesellschaft, Weinheim 1983, pp. 215–226, see p. 222

32. Samuels, A. J., Milby, T. H.: J. Occup. Med. *13* (1971), 147–151
33. Schaller, K. H., Angerer, J., Lehnert, G.: Bio-Monitoring in der Arbeits- und Umweltmedizin. Deutsches Ärzteblatt – Ärztliche Mitteilungen, Heft 31/32, August 1993
34. Sharbrough, F. W. in: Niedermeyer, E., Lopes da Silva, F. (eds.): Electroencephalography. 2nd ed. Urban & Schwarzenberg, Baltimore 1987, pp. 163–181
35. Tusell, J. M., Suñol, C., Gelpi, E., Rodriguez-Farré, E.: Arch. Toxicol. *60* (1987), 432–437
36. Ulmann, E.: Lindan. Schillinger, Freiburg/Br. 1973
37. Vohland, H. W., Koransky, W. in: Deutsche Forschungsgemeinschaft (ed.): Hexachlorcyclohexan als Schadstoff in Lebensmitteln. Forschungsbericht. VCH Verlagsgesellschaft, Weinheim 1983, pp. 246–262
38. Vohland, H. W., Portig, J., Stein, K.: Toxicol. Appl. Pharmacol. *57* (1981), 425–438
39. Woo, Y.-T., Lai, D. Y., Arcos, J. C., Argus, M. F.: Chemical Induction of Cancer. Vol. 3 B: Aliphatic and Polyhalogenated Carcinogens. Academic Press, Orlando 1985, p. 188
40. World Health Organization (WHO): Recommended health-based limits in occupational exposure to pesticides. WHO technical report series No. 677. WHO, Geneva 1982, pp. 57–85
41. Zielhuis, R. L., Verberk, M. M.: Int. Arch. Arbeitsmed. *32* (1974), 167–190

Authors: N. Wichmann, B. Eistel
Approved by the Working Group: 29th September 1988

Nitrobenzene

BAT	100 µg aniline/l blood set free from haemoglobin adduct Sampling time: with long-term exposure: after several consecutive shifts
Date of evaluation	1986
Formula	$C_6H_5NO_2$

Molecular weight	123.11
Melting point	5.7 °C
Boiling point	210.8 °C (760 torr)
Solubility in water	1.9 g/l (20 °C)
MAK [last established: 1958]	1 ml/m^3 (ppm) 5 mg/m^3

1 Metabolism and Kinetics

For occupational exposure to nitrobenzene uptake via both the lungs and the skin is important (18). According to the literature, after exposure at the level of the current MAK value on average up to 35 mg of nitrobenzene per workshift can be absorbed (2, 15).

Nitrobenzene is, in a reductive metabolic step, transformed first of all into nitrosobenzene (12, 22). In this metabolic step intestinal bacteria seem to be involved (16). Furthermore, nitroreductases have been detected in the liver (21). Nitrosobenzene is transformed in a subsequent step in the liver to phenylhydroxylamine; this can also take place in erythrocytes. The reverse reaction in which phenylhydroxylamine is converted to nitrosobenzene in erythrocytes takes place in a coupled reaction whereby methaemoglobin is formed from haemoglobin.

Nitrosobenzene, formed as an intermediate in the metabolism of nitrobenzene, is chemically reactive and can react with glutathione and SH-groups of proteins containing cysteine to form a glutathione conjugate or a protein conjugate (1, 5, 6, 19).

An example of the protein adduct formation is the haemoglobin conjugate. It forms the basis of the present BAT value evaluation. The metabolic pathways of nitrobenzene are summarised in figure 1. For further details, reference is made to the paper setting out basic principles published by the Commission (4) and the BAT documentation for aniline (10).

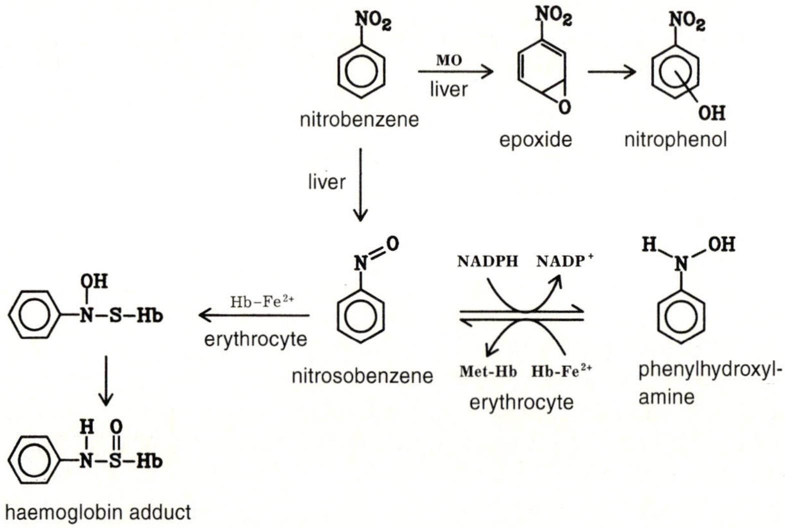

Fig. 1: Metabolism of nitrobenzene in hepatocytes and erythrocytes (simplified). MO = monooxygenase, $Hb\text{-}Fe^{2+}$ = haemoglobin, Met-Hb = methaemoglobin. For further details reference is made to the BAT value documentation for aniline (10).

In man and animals, nitrobenzene metabolism is considerably more protracted than aniline metabolism; it leads however to comparable methaemoglobin levels (1). Most of the nitrobenzene absorbed is metabolised in humans to *p*-nitrophenol and to *p*-aminophenol. A background level of about 25 µg *p*-nitrophenol/l and of about 1000 µg *p*-aminophenol/l urine must be taken into account in the evaluation (3, 13).

The kinetics of these metabolic processes can be influenced in the individual by a number of factors. In persons with congenital enzyme defects, such as a lack of glucose-6-phosphate dehydrogenase, weak acetylator status and haemoglobin anomalies, an increased sensitivity to nitrobenzene is to be expected (8, 11).

Under anaerobic conditions, the reduction of nitrobenzene also produces free nitrobenzene-anion radicals which either disproportionate into nitrobenzene and nitrosobenzene or, in the presence of oxygen, can react to yield nitrobenzene and H_2O_2 or with proteins or DNA to form conjugates (27, 28). The oxygen radicals or H_2O_2 formed are held responsible for the cytotoxic properties of nitrobenzene (12).

2 Critical Toxicity

After intake of nitrobenzene, with a certain latency period, cyanosis develops due to the formation of methaemoglobin. After higher level nitrobenzene exposures, depending upon the individual, nausea, headaches, drowsiness and even coma have been observed. Rapid uptake of larger amounts of nitrobenzene can result immediately in nausea, vomiting, headaches, paralysis and unconsciousness. Chronic intoxication can lead to haemolysis, liver damage and in rare cases also to cutaneous efflorescences (3, 13, 23). The lowest lethal dose that has been reported for humans was 35 mg/kg (17).

The production of methaemoglobin can lead to the formation of Heinz bodies in the erythrocytes. Neurotoxic effects have been described which in some cases were said to be more severe than after comparable aniline exposure (18). In animal experiments nystagmus, paralysis, oversensitivity, tremor and the loss of reflexes have been observed (20). After sub-acute and chronic exposure various authors have reported liver damage in humans and experimental animals (3).

So far no conspicuous signs of carcinogenic or teratogenic potential of nitrobenzene have been observed in either animal studies or human case studies (3). No clear correlations can be recognised in the reproductive effects reported in connection with nitrobenzene (3, 24). In the Ames test no significant mutagenic activity of nitrobenzene could be detected (3, 25), although it can bind to DNA under anaerobic conditions (26). Reports of genotoxic effects of some metabolites of nitrobenzene such as *p*-nitrophenol, *p*-aminophenol, *p*-nitrosophenol, phenylhydroxylamine, acetanilide or aniline are inconsistent. Further investigations are necessary for clarification of this point (3, 29).

3 Exposure and Effects

The acute toxicity of nitrobenzene is to be attributed to the production of methaemoglobin which results from the metabolic formation of the intermediates nitrosobenzene and phenylhydroxylamine. Normally a methaemoglobin value of up to 5 % is regarded as tolerable. This has been discussed in detail elsewhere (4).

In persons chronically exposed to nitrobenzene the classic symptoms are characterised as follows: fatigue, lack of appetite, general stomach complaints, weakness, dizziness, depression, at a later stage anaemia, liver function disorders, Heinz bodies, disorders of kidney function.

Nitrobenzene exposure levels can be determined even in the non-toxic range with the biochemical effect parameter of protein adduct formation. The determination of the aniline haemoglobin conjugate proved to be suitable for this purpose (1, 11, 14), whereby the aniline set free from the adduct by hydrolysis serves as the parameter (11).

4 Selection of the Indicators

Internal nitrobenzene exposure can be determined by measuring the concentration of nitrobenzene, aniline, nitrophenol or aminophenol or their conjugates in whole blood and urine.

Determination of the methaemoglobin level is diagnostically important for evaluating the severity of an acute intoxication and continual adjustment of the dose of antidote. Methaemoglobin determination is therefore indispensable in the treatment of acute nitrobenzene intoxications.

For the evaluation of chronic or intermittent and low-level nitrobenzene exposures, methaemoglobin formation is less suitable as a parameter of effect as rapid reversibility is observed and thus the time of sampling has a great influence on the results. These difficulties are avoided if the formation of the "aniline haemoglobin adduct" which is formed from nitrosobenzene and haemoglobin is used. The amount of the toxic intermediate, nitrosobenzene, formed metabolically is represented in the level of this parameter.

The haemoglobin conjugate of aniline can be detected even after days after exposure and even when the causal nitrobenzene and its metabolites have already been completely eliminated. Unlike the determination of methaemoglobin, the results of the determination of this conjugate are almost independent of the sampling time (4).

5 Methodology

The aniline haemoglobin conjugate is isolated from a blood sample and the aniline set free after hydrolysis is quantitatively determined (11).

6 Evaluation of the BAT Value

In laboratory investigations with volunteers exposed to nitrobenzene by inhalation of increasing amounts, the correlating methaemoglobin values in blood and the aminophenol and nitrophenol excretion rates in the urine were determined without a tolerable threshold limit value being formulated (9,13,15). In persons not exposed, on average methaemoglobin values of 1.2 ± 0.81 % are found (7).

To date, a maximum permissible methaemoglobin value of 5 % has usually been applied for persons handling nitrobenzene (4).

For the reasons presented in the BAT documentation for aniline, however, for the setting of a BAT value the determination of the aniline haemoglobin conjugate is preferred.

For the determination of a BAT value recourse has to be made to as yet unpublished results obtained by the Commission in the elaboration of biological monitoring procedures for aromatic amines (11). In these investigations results are available for 250 persons chronically exposed to nitrobenzene; in addition 20 cases of acute nitrobenzene intoxication have been evaluated. The result of this investigation was that at methaemoglobin levels up to 5 %, less than 100 µg of aniline is always set free from the haemoglobin conjugate per litre of whole blood. Further studies in support of this concept are urgently necessary. When the results of such studies are available, the BAT value established provisionally here will be modified if necessary.

On the basis of the data mentioned above a BAT value for nitrobenzene is set as an

aniline content (released aniline) in the isolated haemoglobin adduct of 100 µg of aniline per l of whole blood.

Observation of this value of the parameter of effect can, as far as we know at present, prevent damage attributable to a restriction of the oxygen transport capacity of the erythrocytes.

7 Interpretation of the Data

In analogy to the principles presented in the BAT documentation for aniline, the aniline haemoglobin adduct can be measured after at least 3 consecutive working days at any time during or after the shift.

Reference should be made here to the fact that as for aniline, in principle also the excretion of free aniline in the urine can be measured. The BAT value for the free aniline in urine is 1 mg/l. From a diagnostic point of view, however, the aniline released from the haemoglobin adduct is to be preferred. For methodological reasons, determination of free aniline in urine may be justified in some cases.

Reference should be made to the BG-Grundsatz G 33 (guideline issued by the Employers' Liability Insurance Association) (8) for the deadlines for follow-up examinations for occupational medical health surveillance.

8 References

1. Albrecht, W., Neumann, H. G.: Biomonitoring of aniline and nitrobenzene. Hemoglobin binding in rats and analysis of adducts, Arch.Toxicol. *57* (1985), 1–5
2. Baselt, R. C.: Biological monitoring methods for industrial chemicals. Davis: Biomedical Publ. (1980), 206–207
3. Beauchamp Jr., R. O., Irons, R. D., Rickert, D. E., Couch, D. B., Hamm Jr., T. E.: A critical review of the literature on nitrobenzene toxicity. CRC Crit. Rev. Toxicol. *11* (1982), 33–84
4. Bolt, H. M., Neumann, H. G., Lewalter, J.: Zur Problematik von BAT-Werten für aromatische Amine. Arbmed. Sozmed. Prävmed. *20* (1985), 197–201
5. Eyer, P.: Reactions of Nitrobenzene with reduced Glutathione. Chem. Biol. Interact. *24* (1979), 227–239
6. Eyer, P.: Reactions of Nitrosoarenes with Sulphydryl Groups: Reaction Mechanism and Biological Significance. In: Gorrod/Damani (eds.): Biological oxidation of nitrogen in organic molecules. Ellis Horwood, Chichester 1985), pp. 386–399
7. Gobbi, A., Soverini, R., Grisler, R.: Valov Normalidi Methemoglobinemia rilevati in 500 Abitani di Milano, Med. Lav. *65* (1974), 307–310

8. Hauptverband der gewerblichen Berufsgenossenschaften: Berufsgenossenschaftliche Grundsätze für arbeitsmedizinische Vorsorgeuntersuchungen: G 33 Aromatische Nitro- oder Aminoverbindungen, 2.A., Gentner Verlag, Stuttgart 1981
9. Ikeda, M., Kita, A.: Excretion of *p*-nitrophenol and *p*-aminophenol in the urine of a patient exposed to nitrobenzene. Br. J. Ind. Med. *21* (1964), 210–213
10. Lewalter, J.: Anilin. In: Henschler, D. (ed.): Biologische Arbeitsstoff-Toleranz-Werte (BAT-Werte) und Expositionsäquivaltente für krebserzeugende Arbeitsstoffe (EKA). Arbeitsmedizinisch-toxikologische Begründungen, 3rd issue. VCH Verlagsgesellschaft, Weinheim 1986
11. Lewalter, J., Korallus, U.: Blood protein conjugates and acetylation of aromatic amines; New findings on biological monitoring. Int. Arch. Occup. Environ. Health *56* (1985), 179–196
12. Mason, R. P.: Free radical metabolites of foreign compounds and their toxicological significance. In: Hodgson/Bend/Philpot (eds.): Reviews in biochemical toxicology. Elsevier, New York 1979, pp. 151–200
13. Myslak, Z., Piotrowski, J. K., Mussialowicz, E.: Acute nitrobenzene poisoning. A case report with data on urinary excretion of *p*-nitrophenol and *p*-aminophenol. Arch. Toxicol. *28* (1971), 208–213
14. Neumann, H. G.: Analysis of hemoglobin as a dose monitor for alkylating and arylating agents. Arch. Toxicol. *56* (1984), 1–6
15. Piotrowski, J.: Further investigations on the evaluation of exposure to nitrobenzene. Br. J. Ind. Med. *24* (1967), 60–65
16. Reddy, B. G., Pohl, T. R., Krishna, G.: The requirement of gut flora in nitrobenzene-induced methemoglobinemia in rats. Biochem. Pharmacol. *25* (1976), 1119–1122
17. Sax, N. I.: Dangerous properties of industrial materials, 6th ed. Van Nostrand Reinhold, New York 1984, p. 2010
18. Seeger, R., Neumann, H. G.: Nitrobenzol. Dt. Apoth. Z. *126* (1986), 597–598
19. Skipper, P. L., Green, L. C., Bryant, M. S., Tannenbaum, S. R., Kadlubar, F. F. in: International Agency for Research on Cancer (IARC): Monitoring human exposure to carcinogenic and mutagenic agents. IARC Scientific Publications No. 59. IARC, Lyon 1984, pp. 143–150
20. Smith, R. P., Alkaitis, A. A., Shafer, P. R.: Chemically induced methemoglobinemia in the mouse. Biochem. Pharmacol. *16* (1967), 317–328
21. Uehleke, H.: Nitrosobenzol und Phenylhydroxylamin als Zwischenstoffe der biologischen Reduktion von Nitrobenzol. Naturwissenschaften *50* (1963), 335–336
22. Uehleke, H.: Nitrosobenzol im Blut von Katzen nach Verabreichung von Nitrobenzol. Arch. Exp. Path. Pharmacol. *247* (1964), 412–418
23. Wirth, W., Gloxhuber, C.: Toxikologie. 3rd ed. Thieme Verlag, Stuttgart 1981, pp. 231–232
24. Hatakeyama, S., Kovacs, K., Jeghiayan, E., Blaschek, J. A.: Aniline-induced changes in the corpora lutea of rats. Am. J. Obst. Gynecol. *109* (1971) 469–473
25. Chiu, C. W., Lee, L. H., Wang, C. Y., Bryan, G. T.: Mutagenicity of some commercially available nitro compounds for Salmonella typhimurium. Mutat. Res. *58* (1978) 11–20
26. Greenstock, C. L., Chapman, J. D., Raleigh, J. P., Shierman, E., Reuvers, A. P.: Competitive radioprotection and radio-sensitization in chemical systems. Radiat. Res. *59* (1974) 556–560
27. Josephy, P. D., Mason, R. P.: Nitroimidazoles. In: Anders, M. W. (ed.): Bio-activation of Foreign Compounds. Academic Press, New York 1985, pp. 451–483.
28. Boyd, M. R.: Biochemical mechanisms in chemical induced lung injury: Roles of metabolic activation. CRC, Critical Reviews in Toxicology (1980), 103–176
29. McCoy, E. C.: Role of metabolism on the mutagenicity of nitroarenes. In: Greim, H., Jung, R., Kramer, M., Marquardt, H., Oesch, F. (eds.): Biochemical basis of chemical carcinogenesis. Raven Press, New York 1984, pp. 57–67

Author: J. Lewalter
Approved by the Working Group: 21st February 1986

Parathion

BAT

500 µg *p*-nitrophenol/l urine
If exceeded:
Inhibition of the erythrocyte acetylcholinesterase activity:

	Reference value	BAT
Time of determination of activity	Before exposure	After exposure
Value	Individually variable ≙ 100%	Reduction of the activity to 70% of the reference value

Sampling: end of exposure, end of shift

Date of evaluation
1985

Synonyms
E 605
Thiophos®
Paraphos
Nitrostigmine
Folidol®

Chemical name
O,*O*-Diethyl-*O*-(4-nitrophenyl)-monothiophosphate

Formula
$C_{10}H_{14}NO_5PS$

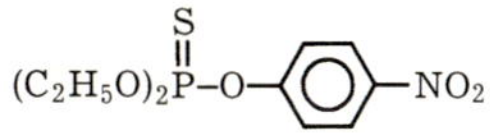

Molecular weight
291.27

Melting point
6.1 °C

Boiling point
375 °C

Vapour pressure at 20 °C
$7.6\ 10^{-4}$ Pa ($0.57\ 10^{-5}$ torr)

MAK
0.1 mg/m^3

Parathion is a highly toxic organophosphate which has been used since about 1949 extensively and in numerous formulations, mainly as an insecticide. Pure parathion and the analogous methyl parathion are yellow oils which are used as vapours, powders and aerosols in up to 50 % concentrations in pesticides. The following account evaluates the exposure parameter for parathion as the urine concentration of *p*-nitrophenol, a relevant metabolite of parathion (and methyl parathion). The corresponding parameter of effect, the erythrocyte acetylcholinesterase activity applies independently for all inhibitors of this enzyme. The reader is referred to the BAT documentation for acetylcholinesterase inhibitors.

1 Metabolism and Kinetics

Parathion can be taken up by inhalation or orally but particularly also through the skin (24). The metabolism of parathion has been investigated in detail and clarified in humans (see Fig. 1), (7, 19, 22, 30). Figure 1 shows the most important metabolic pathways.

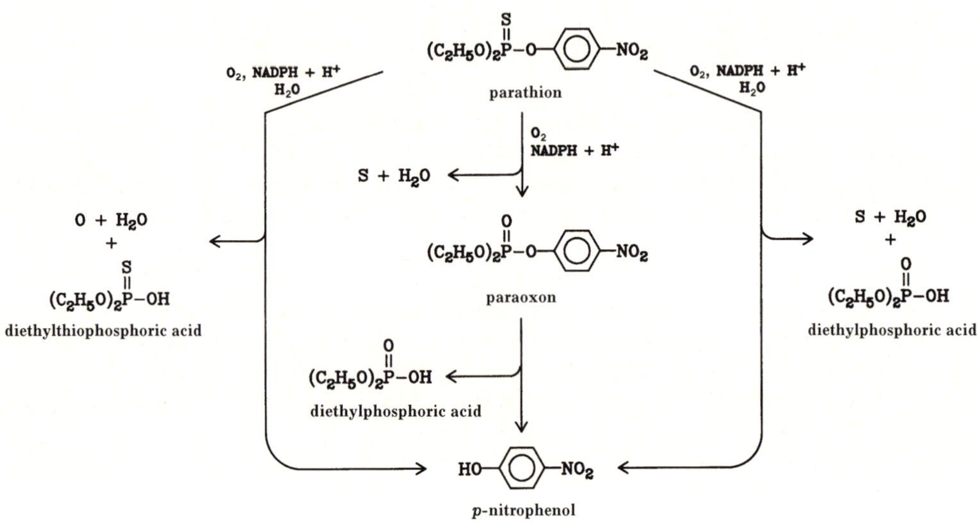

Fig. 1: Metabolism of parathion (adapted from (20))

Parathion can accumulate to a certain extent unchanged in the kidneys, the lungs, but also in the intestinal wall and is subject in these organs to a local metabolism. Most of the incorporated parathion is, however, metabolised in the liver (10). Microsomal enzymes of the liver activate the relatively non-toxic parathion to paraoxon, which is a potent AChE inhibitor. On the other hand, both the unchanged parathion and also the metabolically formed paraoxon are rapidly hydrolysed by plasma and tissue esterases (1), phosphatases, and in particular also by a relatively specific hydrolase, the so-called paraoxonase (11) to form diethylthiophosphoric acid, diethylphosphoric acid and *p*-nitrophenol (5). These products of metabolism are mostly excreted via the urine. They can be considered as indicators of the internal exposure to parathion.

After intravenous injection of radioactively labelled parathion (labelled on the ring and the ethyl group) into Rhesus monkeys, 75 % and 95 % of the injected radioactivity was detected in the urine after 6 and 24 hours, respectively; after 48 hours, 100 % had been recovered (6). Accumulation in the organism could not be detected. Of the excreted parathion metabolites, the phosphoric acid ester is unstable after longer storage so that only the *p*-nitrophenol can be used for exposure monitoring. The *p*-nitrophenol excretion correlates well with the parathion exposure, worse, however, with the corresponding effect (acetylcholinesterase or pseudocholinesterase inhibition). Thus, in persons exposed to parathion with serum concentrations of 4–200 µg/l and excretion of *p*-nitrophenol levels of 0.4–13.2 mg/l the cholinesterase activity in the blood, plasma and erythrocytes was found to be in the normal range (9, 26). After parathion exposures causing more than 30 % inhibition of the erythrocyte acetylcholinesterase and pseudocholinesterases the former regenerate themselves by 1–2 % per day, the latter by 3–9 % per day (12).

2 Critical Toxicity

The clinical picture of parathion intoxication is caused exclusively by the inhibition of the bound acetylcholinesterase at the cholinergenic synapses (27). The acute inhibition of acetylcholinesterase leads to an accumulation of the transmitter substance acetylcholine in the tissues (endogenous acetylcholine intoxication). Depending on the extent of the inhibition, it is accompanied by characteristic clinical signs of parasympathetic stimulation or even by a vegetative, motor and central nervous cholinergenic crisis (16, 28). Details of this syndrome have been described in the BAT documentation Acetylcholinesterase Inhibitors. Inhibition of the acetylcholinesterase (AChE) leads to clinical symptoms only when the AChE activity has decreased markedly below 70 % of the individual reference value (8). With acute parathion intoxication the decrease in the AChE activity to less than 20 % of the reference value is regarded as requiring treatment (4). The directly measurable decrease in erythrocyte AChE activity can be regarded as an indication of critical AChE inhibition in the nervous system (23, 31).

During acute parathion intoxication the extent of the AChE inhibition determines the severity of the observed symptoms (14, 29, 31).

Chronic exposure to parathion can occur during industrial and agricultural contact with the substance. The danger is usually relatively slight as the biological half-life of parathion is less than one hour and that of paraoxon considerably less than seven hours (14). Depending upon the exposure profile and individual pharmacokinetics, however, chronic exposure may lead to intoxication with muscarinic and central nervous symptoms (31). Investigations of persons exposed to parathion and of various species of animal have to date produced no indications of carcinogenic, mutagenic or teratogenic effects (21).

3 Exposure and Effects

The relationship between external exposure (concentration of parathion at the workplace), internal exposure (concentration of parathion or its active metabolite paraoxon, as well as *p*-nitrophenol in the body fluids) and effect (AChE and ChE inhibition) is not only a function of the parathion dose. It is to a great extent also dependent on the individual (genetic) disposition of the exposed person.

Finally in the total evaluation of the individual exposure and effect after contact with parathion it should be borne in mind that this substance is used in agriculture in formulations which also contain other substances. This is dealt with in detail in the BAT documentation Acetylcholinesterase Inhibitors.

4 Selection of the Indicators

The extent of possible, in particular chronic parathion exposures is determined most effectively via the metabolite *p*-nitrophenol. *p*-Nitrophenol can be determined in plasma or urine, whereby the urine assay is easier to carry out. The *p*-nitrophenol formed in the metabolism of parathion is excreted in conjugated form. Accordingly, the determination of the *p*-nitrophenol glucuronide contained in the urine or the *p*-nitrophenol which can be isolated from it after hydrolysis is suitable for the quantitative evaluation of internal parathion exposure for purposes of biological monitoring (17).

During evaluation it should be borne in mind that also in unexposed persons without direct contact with parathion a *p*-nitrophenol body burden up to 100 µg/l urine can be detected. Its sources are still unknown (15). In field studies the *p*-nitrophenol excretion in the urine shows a wide range of variation; about 110 µg/l (60–310 µg/l) in persons resident near orchards sometimes treated with parathion, about 280 µg/l (100–720 µg/l) in workers in the orchards without direct contact with parathion, and about 2.9 mg/l (0.14–11.3 mg/l) urine in the actual parathion users, without an inhibition of the cholinesterases in the blood, plasma or erythrocytes being detected (3). The presence of *p*-nitrophenol in whole blood or plasma of persons exposed to parathion is repeatedly mentioned but the amounts have not been published. According to observations by Commission members apparently *p*-nitrophenol plasma concentrations of about 200 µg/l are tolerated without affecting the erythrocyte AChE or pseudocholinesterase activity.

5 Methodology

The determination of the concentration of *p*-nitrophenol in biological material can take place by chromatographic methods (18). Reliable methods of determination are furthermore described in the loose-leaf collection “Analysen in biologischem Material” (“Analyses of Hazardous Substances in Biological Materials”) of the DFG Commission for the Investigation of Health Hazards of Chemical Compounds in the Work Area (2).

6 Evaluation of the BAT Value

While taking into account the evaluation of the acetylcholinesterase activities described in the BAT documentation "Acetylcholinesterase Inhibitors", in the case of chronic parathion exposure the determination of *p*-nitrophenol in urine is recommended. Despite a variety of investigations on the biological effects of internal parathion exposure and numerous suggestions of tolerance values, so far a critical review of the data is lacking. According to the calculations by Piotrowski (25) theoretically after daily parathion exposure to approx. 0.1 mg/m^3 in the air at the workplace, *p*-nitrophenol levels of the order of magnitude of 500 µg/l urine are to be expected. W. V. Hartwell et al. (13) observed no inhibition of the cholinesterases in blood with *p*-nitrophenol excretion levels of less than 200 µg/l urine, which was confirmed in repeated detailed studies. This is in accordance with the personal experience of Commission members. A BAT value for the *p*-nitrophenol excretion of employees exposed to parathion can, on the basis of the available data, only be set provisionally. With the availability of further information it should be revised if necessary. The BAT value is provisionally set at a urinary excretion level of

500 µg *p*-nitrophenol/l urine

The determination should be carried out in the post-shift urine or immediately after exposure. This BAT value is only valid in connection with the parameter of effect for acetylcholinesterase inhibitors. The inhibition of the acetylcholinesterase activity determined in the erythrocytes must not be greater than 30 %, measured against the individual reference value.

7 Interpretation of the Data

For the reliable determination of representative and individually comparable acetylcholinesterase activities and *p*-nitrophenol excretion, the sampling of blood for the acetylcholinesterase determination and urine collection for the detection of *p*-nitrophenol after 3 consecutive working days has proved suitable. This should take place for all persons present on a shift simultaneously at the shift change. The sampling should be carried out under contamination-free conditions after washing and in private clothes in the medical centre. The evaluated BAT values limit the uptake of parathion through the maximum levels of *p*-nitrophenol excretion in the urine and the corresponding effects in the sense of acetylcholinesterase inhibition. When assessing whether the BAT value has been exceeded the parameter of effect, acetylcholinesterase inhibition, is the critical parameter, not the *p*-nitrophenol excretion.

8 References

1. Aldridge, W. N.: Bull. Wld. Hlth. Org. *44* (1971), 25–30
2. Henschler, D. (ed.): Analytische Methoden zur Prüfung gesundheitsschädlicher Arbeitsstoffe. Band 2: Analysen in biologischem Material. VCH Verlagsgesellschaft, Weinheim 1976
3. Arterberry, J. D., et al.: Arch. Environ. Health *3* (1961), 467–485
4. Barnes, J. M., Davies, D. R.: Brit. Med. J. *2* (1951), 816
5. Comer, S. W., Ruark, H. E., Robbins, A. L.: Bull. Env. Cont. Tox. *16* (1976), 618–625
6. Copeland, F. et al.: Toxicol. Appl. Pharmacol. *19* (1971), 400
7. Crowder, L. A., Whitson, R. S.: Bull. Environ. Contam. Toxicol. *24* (1980), 444–451
8. Dubois, K. P.: Arch. Ind. Health *18* (1958), 488
9. Durham, W. F., Wolfe, H. R., Elliot, J. W.: Arch. Env. Health *24* (1972), 381–387
10. Fredriksson, T., Bigelow, J. W.: Arch. Environ. Health *2* (1961), 663–667
11. Geldmacher-von Malinckrodt, M., Hoppe-Seylers, Z: Physiol. Chem. *354* (1973), 337–340
12. Grob, D., Garlick, W. L., Harvey, A. M.: Bull. Johns Hopkins Hosp. *87* (1950), 106–129
13. Hartwell, W. V., Hayes, Jr., G. R., Funckers, A. J.: Arch. Environ. Health *8* (1964), 820–825
14. Kimmerle, G., Lorke, D.: Pflanzenschutznachrichten Bayer *21* (1968), 111–142
15. Kirby, K. W., et al.: J. Agric. Food Chem. *27* (1979), 757–759
16. Kitz, R. J.: Acta Anaesthesiol. Scand. *8* (1964), 197
17. Lauwerys, R. L.: Industrial Chemical Exposure: Guidelines for Biological Monitoring. Davis: Biomedical Publ. (1983), pp. 118–121
18. Michalke, P.: Z. Rechtsmed. *88* (1982), 195–202
19. Neal, R. A.: Toxicol. Appl. Pharmacol. *23* (1972), 123–130
20. Neal, R. A. in: Anders, M. W. (ed.): Bioactivation of Foreign Compounds. Academic Press, Orlando 1985, pp. 524
21. National Institute for Occupational Safety and Health (NIOSH): Occupational Exposure to Parathion. NIOSH, Cincinnati 1976
22. Norman, J. B., Baughn, W. K., Neal, R. A.: Biochem. Pharmacol. *22* (1973), 1091–1101
23. O'Brien, R. D.: Insecticides: Action and Metabolism. Academic Press, New York 1967
24. Henschler, D. (ed.): Parathion. Gesundheitsschädliche Arbeitsstoffe. Toxikologisch-arbeitsmedizinische Begründung von MAK-Werten. VCH Verlagsgesellschaft, Weinheim 1973/1992
25. Piotrowski, J. K.: Exposure Tests for Organic Compounds in Industrial Toxicology. US Gov. Print. Office, Washington D. C. 1977, pp. 107–111
26. Roan, C. C. et al.: Bull. Env. Cont. Tox. *4* (1969), 362–369
27. Schuh, F. T.: Klinik und Therapie akuter Vergiftungen durch Pflanzenschutz- und Schädlingsbekämpfungsmittel. Oral presentation at the "Zentraleuropäischer Anaesthesie-Kongreß", Bremen 1975
28. Schuh, F. T.: Anaesthesist *25* (1976), 501
29. Valentin, H. et al.: Arbeitsmedizin. Vol. 2: Berufskrankheiten. 2nd ed. Thieme, Stuttgart 1979, pp. 90–93
30. Van den Heede, M. et al.: Med. Fac. Landbouwv. Rijksuniv. Gent *47/1* (1982), 435–446
31. Wirth, W.: Arch. Toxicol *12* (1956), 125–128

Author: J. Lewalter
Approved by the Working Group: 10th May 1985

Tetraethyllead

BAT	25 μg diethyllead/l urine, calculated as Pb 50 μg total lead/l urine
Date of evaluation	1994
Synonyms	lead tetraethyl tetraethylplumbane
Formula	$Pb(C_2H_5)_4$
Molecular weight	323.44
Melting point	–136 °C
Boiling point	87–90 °C/11mm
Vapour pressure at 20 °C	0.2 hPa
MAK [last established: 1994]	50 μg/m^3, calculated as Pb

There are numerous animal experimental investigations on lead alkyl compounds. These are listed in the MAK documentation for tetraethyllead and tetramethyllead. The BAT documentation presented here is restricted mainly to the investigations carried out in humans.

1 Metabolism and Kinetics

1.1 Absorption, Distribution, Elimination

In an inhalation study on volunteers with radioactively labelled TEL* and TML* (^{203}Pb; concentration 1 mg/m^3, 10–40 breaths) 37 % of the inhaled TEL and 51 % of the TML was retained. Within 48 hours 20 % TEL and 40 % TML were exhaled again. The study is only available in the form of an abstract from which the details of the study are not apparent (31).

According to early data, under conditions like those found in practice, 6 % of the absorbed dose of lead alkyls was taken up through the skin (38).

Ingested lead alkyls are absorbed via the gastro-intestinal tract (1). According to

* Abbreviations: TML: tetramethyllead
  TEL: tetraethyllead

Grandjean and Nielsen (29), both tetraalkyllead compounds (TEL, TML) are transformed in part to the corresponding trialkyllead soon after oral uptake by the hydrochloric acid in the stomach.

There are numerous data in the literature on the excretion of metabolites of TEL, as well as on the organ distribution of lead species after acute intoxication with TEL.

In an investigation on the excretion of alkyllead species, excretion products in the urine were determined in a patient intoxicated with TEL on the 21st day after exposure: 50 % of the total amount of lead excreted was in the form of DiEL* and only 2 % was TriEL* (60). This corresponds in principle with later findings on the excretion of TriEL and DiEL in persons occupationally exposed to TEL (56, 8, 62).

The highest concentrations of lead in persons intoxicated with TEL were found in the liver, followed by the kidneys and brain (10, 17, 37).

Results of investigations on lead levels in the urine in cases of intoxication are listed in table 1.

Tab. 1: Lead values in urine (total Pb) after TEL exposure (literature references) (according to BUA, (13) in press)

Exposed persons	Lead concentration (or excretion)	References
3 persons acutely exposed to TEL	400–8000 µg/l	Bolanowska et al., 1967 (10)
13 petrol pump attendants	14.5 µg/24 h	Harmann et al., 1981 (30)
1 worker, TEL refinery	350 µg/l	Nantel et al., 1982 (44)
7 workers, TEL factory	29 µg/24 h	Boscolo and Carmignani, 1988 (11); Carmignani et al., 1988 (14)

1.2 Metabolism

After distribution in the organism TEL is metabolised via oxidative dealkylation to TriEL, DiEL and inorganic lead (16, 20). Oxidative dealkylation is catalysed in the endoplasmatic reticulum of the liver cells by cytochrome P 450 (20, 47, 21). The urine metabolites TriEL and DiEL have, in contrast to TEL, ionic character and therefore their distribution behaviour in the organism is different from that of TEL.

2 Acute Intoxication with Tetraethyllead

The acute lethal dose of organolead compounds for humans is not known exactly but can be estimated. Wiernikowski et al. (58) state an estimated minimum lethal oral dose of 15 ml TEL (approx. 0.35 g/kg body weight) based on 16 accidental or suicidal

* Abbreviations: DiEL: diethyllead
TriEL: triethyllead

cases of intoxication (11 men and 5 women from 16 to 71 years old; observed in Poland 1979–1983). The first symptoms of intoxication are known to occur after ingestion of 6 ml TEL (approx. 0.14 g/kg body weight). Grandjean (28) estimates a LD_{50} value of 0.25 g TEL/kg body weight for humans.

The first intoxications with TEL were reported only a few months after the start of commercial production in October 1924 (41). By 1984 150 cases of intoxication with TEL with a lethal outcome were known. Cases of intoxication during cleaning of petrol tanks have also been described (28).

The latency period between TEL exposure and the occurrence of first symptoms ranges from a few hours in severe cases to approx. 10 days (52). Signs of intoxication described include loss of appetite, nausea and vomiting, sleeplessness, weakness, headache, aggression, depression, irritability, diffuse pains, tremor, hyperactivity, increased proprioceptive muscle reflexes, muscle weakness, concentration and memory difficulties, confusion and unusual sensory sensations such as e.g. metallic taste in the mouth (3, 6, 17, 32, 34, 36, 43, 45, 52).

If large amounts of TEL are taken in orally gastro-intestinal symptoms are predominant (vomiting, abdominal pains, diarrhoea) (55).

From the occurrence of the first symptoms it can take hours or even days until the condition of the patient becomes critical. Acute psychosis, convulsions, delirium, fever and coma then develop (17, 59). Further symptoms accompanying organolead intoxication include hypotonia and bradycardia (5, 17, 36, 41). The longer the symptom-free phase after acute intoxication lasts, the better the prognosis. Even after very severe clinical symptoms practically complete recovery is supposed to have taken place within 2-6 months (52).

Autopsy of intoxicated persons revealed degenerative changes of nerve cells of Auerbach's plexus and Meissner's plexus and in the CNS reduction of the neurons in the cortex, coagulation necrosis of the ganglions and myelinoclastic changes in the medulla oblongata and spongial transformation of the white matter (25).

3 Chronic Effects of Tetraethyllead

In the last 20 years intoxication in "petrol sniffers" has been repeatedly reported in the international literature (35, 51, 19, 24, 23).

As a result of such chronic exposure excitability, depression, headache, low blood pressure, neurasthenia, tremor, ataxia, chorca and hallucinations occur. Patients, who often sniffed petrol for many years from an early age, were often mentally greatly retarded (15, 61, 46, 48, 54, 26, 27, 12, 19, 22). With increased blood lead values abnormal EEG findings were recorded (18, 40). Autopsy of petrol sniffers who had died showed brain oedema, necrosis of nerve cells in the hippocampus, loss of Purkinje's cells in the cerebellum and demyelination of the pons (48, 57), as well as atrophy of the cerebellum and altogether a reduced brain weight (33).

Robinson (49, 50) carried out investigations on the mortality and the general state of health of TEL workers who had handled TEL for at least 20 years. In a mortality study (50) 592 workers exposed to TEL (51 deaths) were compared with 660 non-exposed workers (68 deaths). In a further cohort study on the state of health (49) 153 workers exposed to TEL and 153 non-exposed workers were investigated. Examinations were carried out and data collected on height and weight, systolic and diastolic blood-pressure, haemoglobin content of the blood, electrocardiogram, number of children and days sick off work. With regard to the state of health and mortality, between the workers exposed to TEL and the non-exposed control groups in these studies no significant differences were found. The meaningfulness of these studies were, however, called into question later by Grandjean and Nielsen (29). Their criticism was that at the beginning of the mortality study 89 % of the TEL workers were still under the age of 40. Furthermore, the medical examination did not determine any specific long-term effects such as cerebral functions; moreover, no worker was investigated who had in the meantime left the firm.

In 1595 male workers who were employed in a refinery between 1949 and 1982 increased mortality due to lung, kidney and brain tumours as well as leukaemia was determined. To what extent TEL was involved in the induction of tumours could not be deduced from these data as exposures to mixtures of hydrocarbons, hydrogen sulphide and other substances were involved (4).

Further, investigations on 38 employees in a German factory producing lead alkyls have been described (9, 2, 53). The exposure to tetraethyllead was considerably greater than to tetramethyllead. The average tetraethyllead exposure varied individually between 0.6 and 43.1 $\mu g/m^3$ (calculated as Pb). In this study extensive neuropsychological investigations were carried out in which individual correlations between exposure indicators and behavioural parameters were apparent. In the course of the investigation these proved, however, to be inconstant and thus could not be validated (53). Therefore, the range of exposure described in this study can be considered as being below a no-effect level with regard to the occurrence of CNS effects.

Zhang et al. (62) reported on changes in some clinical parameters in a study in the province Hubei/China in which 277 petrol depot workers exposed to tetraethyllead, 36 municipal traffic police and 342 public office clerks were investigated. The average tetraethyllead exposures ($\pm s_x$) in these groups were 1.1 ± 0.4 (office employees), 5.2 ± 2.4 (traffic police) and 84.8 ± 34.3 $\mu g/m^3$ tetraethyllead (calculated as lead). No conspicuous changes were found in the traffic police. There was a significantly increased occurrence of tremor (hands, tongue, eye-lids) in the employees at the petrol depot, which was recognisable above exposure levels of 50 $\mu g/m^3$ (as lead) and intensified with greater exposures in a dose-dependent manner. Furthermore, sinus bradycardia was seen more often in the petrol depot employees than in the office workers. As differences in the physical activity and fitness between these two groups must be taken into consideration, this finding is difficult to interpret. Within the group of petrol depot workers a clear tendency towards a dose-dependent increase in such changes with the level of tetraethyllead exposure was, however, once again conspicuous.

4 Methodology

In the past, in order to monitor persons exposed to lead alkyls, determination of Pb in blood or urine was carried out using AAS. Such methods (Pb in biological material) have been evaluated and published by the DFG Commission for the Investigation of Health Hazards of Chemical Compounds in the Work Area. They have proved reliable and fulfil the requirements of statistical quality control. However, they have the disadvantage of not being able to differentiate between exposure to inorganic and organic lead as the lead species are not analysed separately.

Turkakiewicz and Chmielnicka (56) isolated diethyllead from the urine of a small collective of 26 persons exposed to tetraethyllead and measured the lead using AAS. The correlation between exposure (TEL in air) and excretion of metabolites (DiEL in urine) was, however, subject to a wide range of scatter and did not allow evaluation of a BAT value.

A more recent method for the determination of diethyllead and triethyllead in urine was described by Blaszkewicz et al. (7, 8). It involves HPLC separation of the lead alkyl species and detection with a chemical reaction detector. This method is used here as a basis for the evaluation of the BAT value.

5 Background Exposures

With the above method for determination of diethyllead in urine, in a collective of German office employees without occupational exposure to lead alkyls, urine levels of 1.4 ± 1.0 µg diethyllead/g creatinine (calculated as Pb) and a total lead level of 2.6 ± 1.7 µg Pb/g creatinine were found (9). Lauwerys and Hoet (39) give a value of 50 µg/g creatinine as an upper limit of normal total lead excretion. However, more recent results indicate that this upper limit is now probably considerably lower because of the general decrease in environmental lead levels. At present a value of 20 µg/l urine is given as the upper limit in urine samples from persons not occupationally exposed to lead (42).

With the same method (7, 8) in investigations in China on office workers without occupational Pb contact, excretion of diethyllead at a level of 3.8 ± 1.8 µg/l urine (as Pb) and of total lead at a level of 16.9 ± 9.8 µg Pb/l urine were determined. The higher background exposure of these persons compared to persons living under Western European conditions was attributed to a higher exposure to street traffic emissions in the People's Republic of China (62).

6 Evaluation of the BAT Value

According to more recent studies the excretion of the specific TEL metabolite, diethyllead, can be regarded as a suitable parameter for the evaluation of internal TEL exposure (56).

In the study mentioned above on 38 employees in a German factory producing lead alkyls (53) individual excretion levels of diethyllead in urine in the range between 1.22 and 17.25 ($\bar{x}$ ± S.D.: 5.55 ± 4.26) (µg/g creatinine (calculated as Pb) and of total lead between 6.79 and 45.13 ($\bar{x}$ ± S.D.: 21.42 + 10.97) µg Pb/g creatinine were found. The blood lead levels were between 70 and 430 ($\bar{x}$ ± S.D.: 216 ± 75) µg/l. In this investigation, neurological test procedures revealed no constant effects on behaviour. It can be concluded from this result that the range of exposure covered by these values is below a no-effect threshold for central nervous effects.

In the already mentioned new study on employees exposed to tetraethyllead in the People's Republic of China (62), in persons who were exposed to more than 50 µg TEL/m^3 (as Pb), that is, in persons with urine DiEL levels above 20 µg/l (as Pb), the reported occurrence of tremor increased in a dose-dependent manner.

The MAK value for TEL was re-set in 1994 at 50 µg/m^3 (calculated as Pb) (see MAK documentation for tetraethyllead).

The relationships between external occupational exposure to TEL and urinary excretion of TriEL, DiEL and total Pb which were revealed in the study by Zhang et al. (62), are shown in figures 1–3. Comparison of figure 1 (DiEL in urine) with figure 2 (TriEL in urine) shows that DiEL, due to its much greater excretion rate, is the most suitable parameter for biological monitoring.

A urinary level of approx. 25 µg diethyllead (as Pb) per litre is consistent with the MAK value for TEL of 50 µg/m^3.

The BAT value has therefore been set at

25 µg diethyllead/l in urine (calculated as Pb).

The total excretion of lead in urine (Fig. 3) which was determined by Zhang et al. (62) in persons exposed in China correlates only unsatisfactorily with the current occupational TEL exposure. This could be due, on the one hand, to earlier exposure to Pb compounds and the resulting accumulation of Pb in the organism, or to exposures to mixtures of organic and inorganic lead. The latter situation is probable in the collective of Chinese traffic police included in the study; they were exposed on average to only about 5 µg/m^3 TEL but had markedly increased total lead excretion levels (see second point on the graph in Fig. 3). This clearly shows the limitations of the use of the parameter total Pb excretion in persons exposed to lead alkyls.

The BAT value evaluated above on the basis of the DiEL excretion has the disadvantage of requiring use of a complicated method of determination which has not yet been officially validated. An additional BAT value for total Pb excretion would therefore seem desirable for orientation in practice. On the basis of the above-mentioned investigations by Seeber et al. (53) and Zhang et al. (62) such a value, which corresponds to a TEL exposure at the level of the MAK value and a DiEL excretion at the level of the BAT value should be set at approx. 50 µg Pb/l urine.

The BAT value for the total lead excretion is therefore set at

50 µg Pb/l urine.

When interpreting results it should be taken into account that this parameter is less specific.

In practice in persons exposed to TEL, total Pb in urine should be determined routinely because the assay is simpler to carry out. In cases where the urine lead level reaches or exceeds 50 µg/l, specific determination of the urine DiEL level is necessary to establish whether the BAT value has really been exceeded.

Attention is drawn to the necessity of observing the BAT value for lead (Pb concentration in blood) whenever there is exposure to lead. This also applies to chronic exposure to lead alkyls as part of the incorporated lead alkyl is dealkylated to inorganic lead and this accumulates very markedly due to its storage in the bones.

7 Interpretation of the Data

Reference is made to section 6 in particular for the evaluation of the results of lead determination in urine and blood of persons exposed occupationally to tetraethyllead.

8 References

1. Akatsuka, K.: Tetraalkyl lead poisoning. Sangyo Igaku *15* (1973), 3–66
2. Backes, U., Bibernell, U., Neidhart, B.: Air sampling procedure for tetraalkyllead compounds. Fresenius Z. Anal. Chem. *333* (1989), 706
3. Beattie, A. D., Morre, M. R., Goldberg, A.: Tetraethyl-lead poisoning. Lancet *2* (1972), 12–15
4. Bertazzi, P. A., Pesatori, A. C., Zocchetti, C., Latocca, R.: Mortality study of cancer risk among oil refinery workers. Int. Arch. Occup. Environ. Health *61* (1989), 261–270
5. Bhansali, A. B.: Medical aspects of tetraethyl lead operations in a petroleum refinery. Ind. J. Occup. Health *1* (1958), 3–7
6. Bini, L., Bollea, G.: Fatal poisoning by lead-benzine. J. Neuropath. Exp. Neurol. *6* (1947), 271–278
7. Blaszkewicz, M., Baumhoer, G., Neidhart, B.: Kopplung von HPLC und chemischen Reaktionsdetektor zur Trennung und Bestimmung von bleiorganischen Verbindungen. Fresenius Z. Anal. Chem. *317* (1984), 221–225
8. Blaszkewicz, M., Baumhoer, G., Neidhart, B.: Bestimmung von Trialkylbleispezies in Urin. Fresenius Z. Anal. Chem. *325* (1986), 129–135
9. Blaszkewicz, M., Baumhoer, G., Neidhart, B., Myslak, Z., Rest, T., Wandel, E.: Untersuchungen zur berufsbedingten Belastung durch Tetraalkylblei: Arbeitsmedizinische und analytische Aspekte. Verh. Dt. Ges. Arbeitsmed. *28* (1988), 603–605
10. Bolanowska, W.: Distribution and excretion of triethyllead in rats. Brit. J. Ind. Med. *25* (1968), 203–208
11. Boscolo, P., Carmignani, M.: Neurohumoral blood pressure regulation in lead exposure. Environ. Health Perspect. *78* (1988), 101–106
12. Brown, A.: Petrol sniffing lead encephalopathy. N. Z. Med. J. *96* (1983), 421–422
13. BUA, Beratergremium für umweltrelevante Altstoffe: TEL und TML. S. Hirzel Verlagsgesellschaft (in press).
14. Carmignani, M., Boscolo, P., Brandi, G.: Plasma renin activity in occupational exposure. Acta Med. Rom. *26* (1988), 277–286
15. Carroll, H. G., Abel, G. G.: Chronic gasoline inhalation. South Med. J. *66* (1973), 1429–1430
16. Casida, J. E., Kimmel, E. C., Holm, B., Widmark, G.: Oxidative dealkylation of tetra-, tri-, and dialkyltins and tetra- and trialkylleads by liver microsomes. Acta Chem. Scand. *25* (1971), 1497–1499
17. Cassells, D. A. K., Dodds, E. C.: Tetra-ethyl lead poisoning. Brit. Med. J. *9* (1946), 681–685
18. Corsi, G. C., Picotti, G.: Analisi clinica su una casuistica di intossicati da piombo tetraetile. Folia Med. *48* (1965), 856–887

19. Coulehan, J. L., Hirsch, W., Brillman, J., Sanandria, J. Welty, T. K., Colaiaco, P., Koros, A., Lober, A.: Gasoline sniffing and lead toxicity in Navajo adolescents. Pediatrics *71* (1983), 113–117
20. Cremer, J. E.: Biochemical studies on the toxicity of tetraethyl lead and other organolead compounds. Brit. J. Ind. Med. *16* (1959), 191–199
21. Da Silva, D., Diehl, H.: Metabolism of tetraorganolead compounds by rat liver microsomal monooxygenase. III. Enzymatic dealkylation of tetramethyl lead compared with tetraethyl lead. Xenobiotica *15* (1985), 789–797
22. Daniels, A. M., Latchman, R. W.: Petrol sniffing and schizophrenia in a pacific island paradise. Lancet *1* (1984), 389
23. Edminster, S. C., Bayer, M. C.: Recreational gasoline sniffing: acute gasoline intoxication and latent organolead poisoning. Case reports and literature review. J. Emerg. Med. *3* (1985), 365–370
24. Fortenberry, J. D.: Gasoline sniffing. Am. J. Med. *79* (1985), 740–744
25. Forycki, Z., Zegarski, W., Bardzik, J.: Two cases of suicidal tetraethyl lead poisoning. Bull. Inst. Marit. Trop. Med. Gdynia *2*(1977), 179–185
26. Gallassi, R., Montagna, P., Pazzaglia, P., Cirignotta, F., Lugaresi, E.: Peripheral neuropathy due to gasoline sniffing. A case report. Eur. Neurol. *19* (1980), 419–421
27. Goldings, A. S., Stewart, R. W.: Organic lead encephalopathy: behavioral change and movement disorder following inhalation. J. Clin. Psychiat. *43* (1982), 70–72
28. Grandjean, P.: Organolead exposures and intoxications. In: Grandjean, P. et al. (eds.): Biological effects of organolead compounds. CRC Press, Boca Raton, Florida, USA 1984, pp. 227–241
29. Grandjean, P., Nielsen, T.: Organolead compounds: environmental health aspects. Residue Rev. *72* (1979), 97–148
30. Harman, A. W., Frewin, D. B., Priestlly, B. G.: Induction of microsomal drug metabolism in man and in the rat by exposure to petroleum. Brit. J. Ind. Med. *38* (1981), 9197
31. Heard M. J., Wells, A. C., Newton, D., Chamberlain, A. C.: Human uptake and metabolism of tetra ethyl and tetra methyl lead vapour labelled with ^{203}Pb. Int. Conf. Manage. Control Heavy Met. Environ.; CEP Ltd., Edingburgh (1979) 103–108; cited in Chem. Abs. 92-88837m
32. Ichok, G.: Tétra-éthyle de plomb. Presse Med. *36* (1928), 1127–1128
33. Kaelan, C., Harper, C., Vieira, B. I.: Acute encephalopathy and death due to petrol sniffing: neuropathological findings. Aust. N. Z. J. Med. *16* (1986), 804–807
34. Kahan, V. L.: Paranoid states occuring in leaded-petrol handlers. J. Mental Sci. *96* (1950), 1043–1047
35. Kaufmann, A.: Gasoline sniffing among children in a pueblo indian village. J. Pediatr. *51* (1973), 1060–1064
36. Kehoe, R. A.: Tetra-ethyl lead poisoning. J. Am. Med. Assoc. *85* (1925), 108–110
37. Kehoe, R. A.: Pharmacology and toxicology of heavy metals: lead. Pharmacol. Ther. Part A, *1* (1979), 161–188
38. Laug, E. P., Kunze, F. M.: The penetration of lead through the skin. J. Ind. Hyg. Toxicol. *30* (1948), 256–259
39. Lauwerys, R., Hoet, P. (eds.): Industrial Chemical Exposure: Guidelines for Biological Monitoring, 2nd ed., Lewis Pub., Boca Raton, USA 1993
40. Law, W. R., Nelson E. R.: Gasoline-sniffing by an adult. J. Am. Med. Assoc. *204* (1968) 144–146
41. Machle, W. F.: Tetra-ethyl lead intoxication and poisoning by related compounds of lead. J. Am. Med. Assoc. *105* (1935), 578–585
42. Merian, E. (ed.): Metals and Their Compounds in the Environment. VCH Verlagsgesellschaft, Weinhein 1991, p. 987
43. Müller, K.: Zur Kausuistik der Bleitetraäthylvergiftung. Zentralbl. Arbeitsmed. Arbeitsschutz *3* (1953), 8–9
44. Nantel, A. J., Weber, J. P., Tatha, C., Benedetti, J. L., Lefebre, M., Bell-Iles, M.: Occupational intoxication by lead-tetra-ethyl: a case presentation and a re-evaluation of the surveillance programs. Vet. hum. toxicol. *24* (1982), 277
45. Norris, C., Gettler, A. O.: Poisoning by tetra-ethyl lead. J. Am. Med. Assoc. *85* (1925), 818–820
46. Procopis, P. G.: Tetraethyl lead poisoning caused by petrol sniffing. Aust. Paediatr. J. *14* (1978), 205
47. Prough, R. A., Stalmach, M. A., Wiebkin, P., Bridges, J. W.: The microsomal metabolism of the organometallic derivates of the group-IV elements, germanium, tin and lead. Biochem. J. *196* (1981), 763–770

48. Robinson, R. O.: Tetraethyl lead poisoning from gasoline sniffing. J. Am. Med. Assoc. *240* (1978), 1373–1374
49. Robinson, T. R.: The health of long service tetraethyl lead workers. J. Occup. Med. *18* (1976), 31–40
50. Robinson, T. R., Rouge, B.: Delta-aminolevulinic acid and lead in urine of lead antiknock workers. Arch. Environ. Health *28* (1974), 133–136
51. Ross, C. A.: Gasoline sniffing and lead encephalopathy. Can. Med. Assoc. J. *127* (1982), 1195–1197
52. Sanders, L. W.: Tetraethyllead intoxication. Arch. Environ. Health *8* (1964), 270–277
53. Seeber, A., Kiesswetter, E., Neidhart, B., Blaszkewicz, M.: Neurobehavioral effects of a long-term exposure to tetraalkyl lead. Neurotoxicol. Teratol. *12* (1990), 653–655
54. Seshia, S. S., Rajani, K. R., Boeckx, R. L., Chow, P. N.: The neurological manifestations of chronic inhalation of leaded gasoline. Develop. Med. Child Neurol. *20* (1978), 323–334
55. Stasik, M., Byczkowska, Z., Szendzikowski, S., Fiedorczuk, Z.: Acute tetraethyl lead poisoning. Arch. Toxicol. *24* (1969), 283–291
56. Turlakiewicz, Z., Chmielnicka, J.: Diethyllead as a specific indicator of occupational exposure to tetraethyl lead. Br. J. Ind. Med. *42* (1985), 682–685
57. Valpey, R., Sumi, S. M., Copass, M. K., Goble, G. J.: Acute and chronic progressive encephalopathy due to gasoline sniffing. Neurology *20* (1978), 507–510
58. Wiernikowski, A., Sadlik, J., Chrostek-Maj, J., Kobylecka, K., Trela, F.: Acute tetraethyllead ($PbEt_4$) poisoning. Folia Med. Cracov *28* (1987), 3–12
59. Yamamura, Y., Takakura, J., Hirayama, F., Yamauchi, H., Yohida, M.: Tetraethyl lead poisoning caused by cleaning work in the aviation fuel tank. Jap. J. Ind. Health *17* (1975), 223–235
60. Yamamura, Y., Yamauchi, H., Arai, F.: Urinary excretion pattern of triethyllead, diethyllead and inorganic lead in the tetraethyl lead poisoning. Ind. Health *19* (1981), 125–132
61. Young, R. S. K., Grzyb, S. E., Crismon, L.: Recurrent cerebellar dysfunction as related to chronic gasoline sniffing in an adolescent girl. Clin. Pediatr. *16* (1977), 706–708
62. Zhang, W., Zhang, G. G., He, H. Z., Bolt, H. M.: Early health effects and biological monitoring in persons occupationally exposed to tetraethyl lead. Int. Arch. Occup. Environ. Hlth. *65* (1994), 395–399

Author: H. M. Bolt
Approved by the Working Group: 5th May 1994

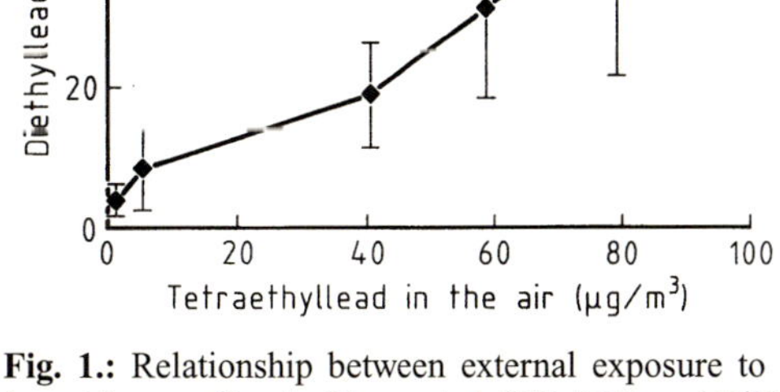

Fig. 1.: Relationship between external exposure to TEL (x-axis) and the excretion of DiEL in urine (y-axis), according to Zhang et al. (62). TEL and DiEL are expressed as the Pb concentration ($\bar{x} \pm$ S.D.)

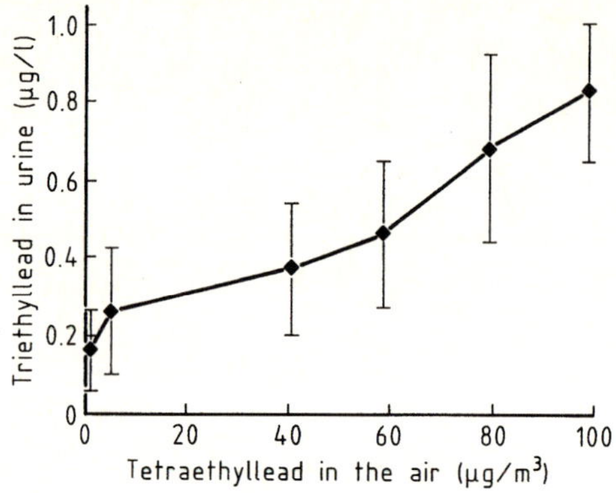

Fig. 2.: Relationship between external exposure to TEL (x-axis) and the excretion of TriEL in urine (y-axis), according to Zhang et al. (62). TEL and TriEL are expressed as the Pb concentration ($\bar{x} \pm$ S.D.)

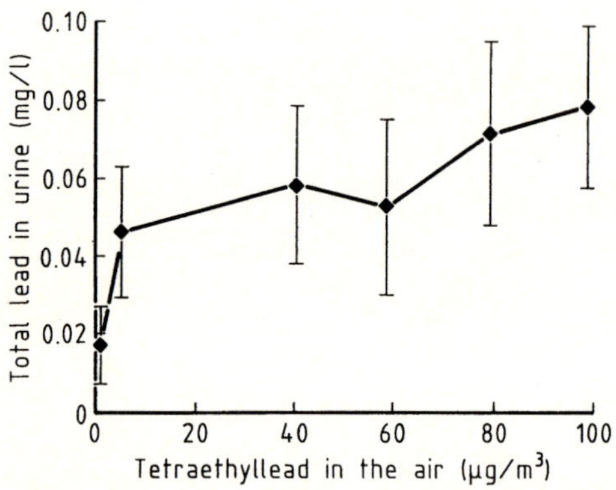

Fig. 3.: Relationship between external exposure to TEL (x-axis) and the excretion of total Pb in urine (y-axis), according to Zhang et al. (62).

Tetrahydrofuran

BAT	8 mg tetrahydrofuran/l urine Sampling time: end of exposure or end of shift
Date of evaluation	1993
Synonyms	Butylene oxide Oxolane Tetramethylene oxide Diethylene oxide
Formula	 C_4H_8O
Molecular weight	72.11
Melting point	–108.5 °C
Boiling point	64–66 °C
Vapour pressure at 20 °C	150 torr
Odour threshold	30 ppm
MAK [last established: 1970]	200 ppm 590 mg/m^3

Introduction

Tetrahydrofuran (THF) is a colourless solvent with an acetone-like smell. The odour threshold is approx. 30 ppm. THF is readily miscible with water and most organic solvents. It is used in many areas, e.g. in adhesives, paints, varnishes, cleaning agents. An important area of application is the production of magnetic tapes. Reactive peroxides can be formed from THF and atmospheric oxygen.

At the workplace THF is taken up mainly as vapour via the respiratory tract. Exposure usually involves mixtures of THF with other organic solvents. THF can also be absorbed via the skin.

Despite its extensive use in industry, effects of THF on health after its inhalation or absorption have so far not been described. The smell of THF is at first pleasant but leads after a short time to disturbance of the sense of well-being. This warning signal generally prevents absorption of large doses during contact with THF.

1 Metabolism and Kinetics

The metabolism of THF in humans and animals has not been studied. Oxidative degradation via 1,4-butanediol to succinic acid is suspected (1). At least part of the absorbed THF is excreted with the urine unchanged.

2 Critical Toxicity

The MAK value documentation which is available for THF dates from 1970 (6). The MAK value, currently 200 ppm, was established to prevent odour irritation. With reference to a study from 1968 the documentation states that concentrations of 500 ppm were considered malodorous by half of the test persons (10). The odour was no longer a nuisance at concentrations under 200 ppm.

In higher concentrations THF, or probably its peroxides formed in the air, irritates mucous membranes and is neurotoxic and hepatotoxic. Central nervous effects have also been observed (7). In humans chronic systemic effects have so far not been reported. To date there are no indications of genotoxic effects of THF.

After very high doses of THF hypersalivation, vomiting, drop in blood pressure with drowsiness, dizziness and collapse, as well as headache, weight loss and albuminuria have been observed. Increase in the muscle tonus is also known. These effects are reversible (2). The liver and kidney damage reported in some cases is probably due less to THF itself than to impurities which have not been characterised. Exposure to THF or to THF together with potential hepatotoxic solvents can lead to increased activity of the liver enzymes in serum (3, 4). Oral uptake of THF can cause haemorrhagic and ulcerative gastritis (8). The changes in the ciliary activity in the trachea seen in animal experiments during exposure to 100 ppm THF have not been observed in humans (8).

More recent investigations of the toxicity of THF in humans have appeared only sporadically. Recently two cases of acute THF intoxication were described (5). Workers who were exposed to an unknown concentration of THF complained of nausea, headaches, impaired vision, pains in the chest, difficulty in breathing and coughing. An increase in the liver enzymes was found in both cases. Other clinical or biological effects were not observed. The symptoms and the disturbances in liver function regressed within two days.

In an investigation into 58 employees working in video tape production and exposed to THF (2 to 150 ppm) no specific effects on health were described (9). It must be added that this was primarily a study on the biological monitoring of THF exposure.

3 Exposure and Effects

With THF the external exposure is defined as any exposure which can lead to absorption of the substance by inhalation or through the skin. The internal exposure is the biologically relevant concentration of THF in the liver, the central nervous system and the olfactory organs. Exposure parameters are the THF concentrations in body fluids such as blood, plasma and urine which correlate with the internal exposure under the specific pharmacokinetic conditions.

3.1 Relationships between External and Internal Exposure

The results of a more recent field study (9) demonstrate the relationship between the external and internal exposure to THF. 58 video tape production workers were investigated who were exposed to 2 to 150 ppm THF. In addition other organic solvents, such as MIBK, toluene and xylohexanone occurred at the workplace.

Figures 1 and 2 contain the correlation diagrams for the relationship between the THF concentrations in the ambient air and those in blood samples and urine samples which were collected after the 8-hour shift. Adjustment of the urine concentrations for density and creatinine did not change the correlation very much; in fact, expressing the concentrations in terms of the creatinine content results in a lower correlation coefficient (see Fig. 2). Sufficiently good correlations were found between the THF concentrations in blood and urine (r between 0.4 and 0.8) (9).

These correlation data were reproduced in a study in which 9 persons were exposed to a THF concentration of about 200 ppm for 8 hours in an exposure chamber (11) (see Tables 1 and 2).

3.2 Relationships between Internal Exposure and Effect

To date there are neither laboratory studies nor field studies available from which relationships between internal exposure and effect can be evaluated.

4 Selection of the Indicators

As described in section 1 (Metabolism and Kinetics) the following can be considered as indicators of internal THF exposure:

- THF in the alveolar air
- THF in blood
- THF in urine

Considerations of practicability lead to selection of the following indicators for the determination of the BAT value:

- THF in blood, and
- THF in urine.

Investigations on non-exposed persons have shown that without occupational exposure no THF can be determined in biological material down to the detection limit of approx. 100 µg/l (9).

5 Methodology

The preferred method of analysis of THF in blood and urine makes use of capillary gas chromatography. The method is described in the literature (9).

6 Evaluation of the BAT Value

The available data provides as a basis for evaluating a BAT value only the relationship between external and internal exposure. The currently valid MAK value of 200 ppm is taken as the permissible external exposure. The evaluation of the MAK value is based, as already mentioned, solely on odour nuisance data from an experimental study. Under these exposure conditions no central nervous or liver toxic effects of THF are to be expected. Results of field studies on employees exposed to THF with regard to the relationships between internal exposure and effects have so far not been published.

A further limitation is that only one field study and one exposure chamber study are available for the evaluation of the BAT value (9, 11). In the field study attention is drawn to the fact that some workers also had intensive skin contact. Moreover the external exposures did not reach the level of the MAK value of 200 ppm.

Regression analysis of the correlation between the THF concentration in blood and urine and the external THF exposure demonstrated that the unmetabolised THF excreted in the urine correlated significantly ($r = 0.86$) with the external THF exposure. The corresponding correlation factor for the analysis of THF in blood was much smaller with $r = 0.50$ (9). The external exposure of 200 ppm THF corresponds to excretion of THF in the post-shift urine at a concentration of approx. 8 mg/l. These correlation data were confirmed in an 8-hour exposure chamber experiment with 9 persons exposed to concentrations in the range of the MAK value for THF (11).

Despite the limited amount of data, it is justified to evaluate a BAT value on this basis because of the relatively good correlation between the THF concentration in the air and that in urine and because of the occupational medical experience of the effects of exposure to such THF concentrations.

Therefore, a BAT value of

8 mg THF/litre urine in the post-shift sample

has been established.

This value should be regarded as provisional. When further data become available the BAT value evaluation should be revised if necessary.

7 Interpretation of the Data

The THF concentration in urine was evaluated without correction for creatinine or density (section 3.1). Important is the time of sampling which should be at the end of the shift, best of all at the end of the working week or shift period.

As THF frequently occurs together with other organic solvents at the workplace further data on the concentration of other components of the mixture are useful. Consideration ought to be given to whether biological monitoring of these substances e.g. methylethyl ketone, toluene, xylenes, is necessary.

The urine samples should be kept cool until analysis, transportation should be carried out in sealed vials (head space vials).

For occupational medical examinations of persons exposed to THF there is at present no Berufsgenossenschaftlicher Grundsatz (guideline issued by the German Employer's Liability Insurance Association). The time intervals for follow-up examinations are oriented on occupational medical requirements; they should, however, in general take place at least every 12 months.

Regression analysis of biomonitoring data revealed that the amount of THF in urine samples collected 16 hours after the end of an 8-hour exposure to THF averaged 2 ± 0.8 mg. As these data differ greatly from the levels in 16-hour urine from the persons exposed to defined THF concentrations in an exposure chamber (see Tab. 2) (11), at present no statement can be made as to the pre-shift value for THF in urine, although this value is important in practice.

It remains to be clarified whether the discrepancies in the pre-shift urine data are to be regarded as the consequences of continual THF exposures which even for a substance with a relatively short biological half-life can lead to an extended urinary excretion via a temporary body burden.

8 References

1. Williams, R. T.: Detoxication mechanisms. 2nd ed., Chapman and Hall, London 1959
2. Hygienic guide series: Tetrahydrofuran. Am. Ind. Hyg. Assoc. J. *20* (1959), 250
3. Juntunen, J., Kaste, M., Harkonen, H.: Cerebral convulsion after enfluran anaesthesia and occupational exposure to tetrahydrofuran. J. Neurol. Neurosurg. Psychiatry *47* (1984), 1258–1259
4. Edling, C.: Interaction between drugs and solvents as a cause of fatty change in the liver. Br. J. Ind. Med. *39* (1982), 198–199
5. Garnier, R., Rosenberg, N., Puissant, J. M. et al.: Tetrahydrofuran poisoning after occupational exposure. Br. J. Ind. Med. *46* (1990), 677–678
6. Henschler, D. (ed.): Tetrahydrofuran. Gesundheitsschädliche Arbeitsstoffe. Toxikologisch-arbeitsmedizinische Begründung von MAK-Werten. Issues 1/15. VCH Verlagsgesellschaft, Weinheim 1972/1990.

7. Elovaara, E., Pfaffli, P., Savolainen, H.: Burden and biochemical effects of extended tetrahydrofuran vapour inhalation of three concentration levels. Acta Pharmacol. Toxicol. *54* (1984), 221–226
8. lkeoka, H., Ohashi, Y., Maruoka, K. et al.: Effects of tetrahydrofuran exposure on the ciliary activity and morphology of tracheal epithelium in rabbits. Osaka City Med. J. *30* (1984), 53–68
9. Ong, C. N., Chia, S. E., Phoon, W. H., Tan, K. T.: Biological monitoring of occupational exposure to tetrahydrofuran. Brit. J. Ind. Med. *48* (1991), 616–621
10. Hofmann, H. Th.: unpublished investigations 1968, Medizinisch-Biologische Forschungslaboratorien – Gewerbehygiene und Toxikologie – der BASF AG, Ludwigshafen
11. Failing A., Knecht, U., Woitowitz, H. J.: Biomonitoring einer standardisierten Tetrahydrofuran-Einwirkung. Daten zur BAT-Wert Festsetzung. In: Kessel, R. (ed.): Verhandlungen der Deutschen Gesellschaft für Arbeitsmedizin und Umweltmedizin e. V., *34* (1994), 375–376, Gentner Verlag, Stuttgart

Author: J. Lewalter
Approved by the Working Group: 5th May 1994

Tab. 1: Average concentrations of THF in blood (mg/l) and alveolar air (ppm) from n = 9 persons, and the standard deviation, after four-hour and eight-hour exposures to concentrations in the range of the MAK value under standardised conditions (11)

Hours after the beginning of exposure	Blood (mg/l)	Alveolar air (ppm)
4	9.6 ± 0.9	20.2 ± 2.9
8	9.8 ± 1.4	21.3 ± 1.3

Tab. 2: Average THF concentrations in urine (mg/l) and the standard deviation during and after standardised THF exposure at concentrations in the range of the MAK value (11)

Hours after the beginning of exposure	THF concentration urine (mg/l)	Number of samples [n]
2	6.2 ± 2.7	5
3	8.6 ± 2.3	4
4	8.0 ± 1.6	5
5	6.2 ± 1.8	3
6	15.0 ± 5.9	2
7	10.6 ± 3.0	3
8	7.9	1
9	7.6 ± 1.1	4
11	6.8 ± 4.0	6
14	1.0 ± 0.7	7
16	0.5 ± 0.2	3
26	0.1 ± 0.1	8
28	0.1 ± 0.1	3
30	0.1	1

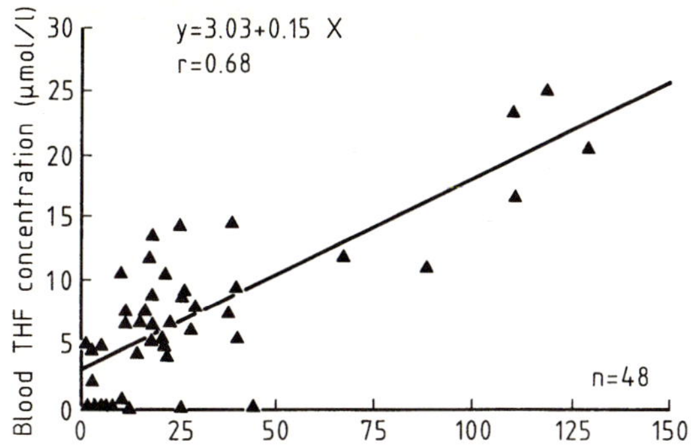

Fig. 1: Relationship between THF concentration in blood and breathing zone air. Points are individual values (9)

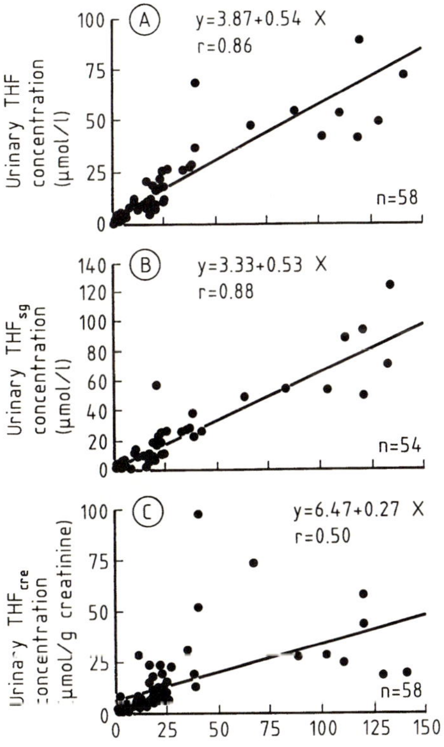

Fig. 2: A = relationship between urinary THF concentration (unadjusted) and breathing zone air.
B = relationship between urinary THF concentration (adjusted for specific gravity of 1.018) and breathing zone air.
C = relationship between urinary THF concentration (µmol/g creatinine) and breathing zone air (according to reference 9)

Xylenes

BAT	2 g toluric acid/l urine 1.5 mg xylene/l blood Sampling time: end of exposure	
Date of evaluation	1984	
Synonyms	Dimethylbenzene	
Formula	$C_6H_4(CH_3)_2$ 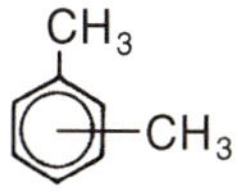	
Molecular weight	106.2	
Melting point	–25 °C (*o*-xylene) –48 °C (*m*-xylene) 13 °C (*p*-xylene)	no data (tech. xylene)
Boiling point	144 °C (*o*-xylene) 139 °C (*m*-xylene) 138 °C (*p*-xylene)	137 °C–140 °C (tech. xylene)
Vapour pressure at 20 °C	6.7 mbar (*o*-xylene) 8 mbar (*m*-xylene) 8.2 mbar (*p*-xylene)	7–9 mbar (tech. xylene)
MAK [last established: 1983]	100 ml/m^3 (ppm) 440 mg/m^3	

1 Metabolism and Kinetics

Due to their high vapour pressure even at room temperature *o*-xylene, *m*-xylene and *p*-xylene are taken up at the workplace predominantly by inhalation. The amount of the toxic substance retained in the body is quoted in the literature as between 60 % and 70 % (1–5). The isomers do not seem to differ in this respect.

Liquid xylene can be absorbed via the intact skin approximately to the same extent as via inhalation during external exposure to 100 ml/m^3 (MAK). Thus, in the literature absorption rates for *m*-xylene of 2 µg and 2.5 µg (cm^2 · min)$^{-1}$ are reported (6,

43). Considerable interindividual fluctuations have been observed (43). On the other hand xylene vapour is absorbed only to a minor extent via the skin (7).

Under workplace conditions the xylene concentration in the blood reaches equilibrium at the latest four hours after the beginning of exposure, which is a feature common to benzene, toluene and xylene aromatics (BTX aromatics) (8). After the end of exposure the concentration of the toxic substance in the blood decreases quickly (3, 9). The half-time for the first elimination phase is given as 30 min. Afterwards the xylene concentration decreases more slowly so that aromatics can be detected in the blood even 16 hours after the end of exposure.

The course of the changes in concentration of the xylenes in the alveolar air with time is more or less parallel to that in blood. After the end of exposure xylene is eliminated in the alveolar air in three elimination phases for which the half-times are given as 0–3 hours, 4–6 hours and >17 hours (9). Following exposure, altogether between 3.5 % and 5.8 % of the retained xylene is eliminated via the lungs (1, 3, 9).

By far the greatest part of the retained xylenes (almost 95 %) is transformed by oxidation of a methyl group into methylated benzoic acids and excreted after conjugation with glycine as methylhippuric acids (toluric acids) in the urine.

Šedivec (1) was able to disprove former assumptions and show that not only the aliphatic side chains of xylenes are oxidised but also the aromatic ring. Approx. 2 % of retained xylenes are excreted in the urine as glucuronides of the corresponding dimethylphenols (xylenols).

The amounts and the concentrations of the toluric acids excreted in the urine increase during exposure and reach their highest values at the end of an 8-hour working period (8). Under workplace conditions the proportion of the total amount of toluric acids excreted which was eliminated during the last two hours of the work shift was between 16 % and 47 % (10). Under experimental conditions in the laboratory this amount varies within narrower limits.

Šedivec and Flek (2) report the amount of toluric acid excreted at the end of the exposure period (8 hours) to be 23 % to 32 %.

2 Critical Toxicity

The literature on the toxicity of the xylenes has recently been summarised and critically reviewed (11–14). The following statements are therefore limited to those effects on health which were observed under exposure conditions within the range of the MAK values. Conditions under which the acute narcotic effects of xylenes occur are not discussed here; instead the reader is referred to the appropriate text books and handbooks of occupational medicine and industrial toxicology (15–18). The present discussion is limited, as far as possible, to observations in humans. Where animal experiments are taken into consideration this is expressly stated. There is no differentiation of the isomeric xylenes as experience of differing toxicity of the individual xylenes in humans is not available and because the results from animal experiments are contradictory on this point.

2.1 Eyes, Skin and Mucous Membranes

Xylene vapours, which can be clearly identified from the smell at concentrations of at least 40 ml/m^3 in the ambient air, begin to irritate the eyes, nose and throat depending upon individual sensitivity between 100 and 200 ml/m^3 (19, 20).

As a consequence of its defatting effect on the intact skin, contact with liquid xylene can lead to dermatitis.

2.2 Central Nervous System

Under conditions of chronic occupational exposure to xylene vegetative-nervous disturbances have been observed more frequently (21–23). According to these reports some of the exposed persons are said to have suffered from vegetative dystonia, headaches, insomnia, irritability, palpitations and digestive disturbances at ambient air concentrations as low as 12 ml/m ^{3}. These studies, accused of having many shortcomings, (13) also suffer from the disadvantage that the complaints described above are neither specific to the substance nor objectifiable with medical scientific methods of investigation.

Recently Finnish authors have attempted to make the characteristic but unspecific health effects of organic solvents on the central nervous system accessible to scientific methods of measurement. Under the defined conditions of exposure possible in laboratory experiments, they subjected up to 22 male test persons to xylene concentrations between 90 and 460 ml/m^3. The test persons were exposed for up to eight days during six hours per day to xylene vapours. The parameters measured included the function of the vestibular organ and other neurophysiological parameters such as reaction time, flicker fusion threshold and manual dexterity. As little as four hours exposure to xylene at a level of 145 ml/m^3 was sufficient to produce impairment of vestibulary function. After a daily six-hour exposure to a xylene concentration of 100 ml/m^3 a decrease in the reaction time was found. After continuous exposure to xylene for only 70 minutes at the same level (100 ml/m^3) Gamberale (24) observed nausea and headaches in some test persons.

2.3 Haematopoietic System

The idea that xylenes, like benzene, cause changes in the blood count has been repeatedly discussed in the literature and is also to be found in text books on the subject (15, 16, 25). This observation was attributed to the presence of benzene which, as little as 25 years ago, was always found as a contaminant of xylene. This description of the differing metabolic behaviour of benzene and its alkyl derivatives was thought to explain the differing effects of these aromatics on the health of human beings. However, more recent reports also indicate that xylene has effects on the blood count (21–23, 26–28). These reports are, however, insufficiently documented (13).

In particular, the more recent finding that also xylenes are oxidised on the aromatic ring, although to a smaller extent than benzene itself, indicate that this question should be given more consideration in future.

Reports of functional disturbances in the liver and kidneys under conditions tolerated at the workplace today are not available. Nor have carcinogenic or teratogenic effects been reported. Animal experiments indicating possible teratogenicity (29–32) suggest that appropriate epidemiological investigations should be carried out on humans (33).

3 Exposure and Effects

As already explained in the section on critical toxicity, above all eyes and mucous membranes, the central nervous system, and possibly the haematopoietic system are the target organs of exposure to xylene.

Under laboratory conditions adverse effects on psychophysiological parameters and on the vestibular organ were found at xylene ambient air concentrations at the level of the currently valid MAK value (100 ml/m^3). For the evaluation of (chronic) occupational exposure, these acute functional disturbances are, however, of limited significance. Even during short-term exposure to substances in laboratory experiments habituation effects were identified. More serious is the suspicion that changes in the blood count occur under workplace conditions at xylene concentrations considerably below the MAK value (21–23).

The expected relationship between the xylene concentrations in the ambient air and the toluric acid excretion in the urine has been observed under laboratory experimental conditions and in field studies. In table 1A the results described in the literature have been listed. Extrapolation to an ambient air concentration of 100 ml/m^3 yields toluric acid concentrations in the urine expected at this concentration of between 0.9 and 3.3 g/l. It should be borne in mind that this is the average value for groups of persons exposed under the same conditions. This averaging conceals the extreme values of the toluric acid concentrations which have been seen in individual samples. Šedivec et al. (2) quote for the individual toluric acid concentration in the urine samples voided at the end of exposure a confidence interval between 1054 and 4018 mg/l. Under workplace conditions this range can be assumed to be even greater (34).

The reasons for this wide range include the differing respiratory minute volumes that occur under workplace conditions (34, 35) and at rest (39). For the same reasons under identical conditions of exposure between 20 % and 25 % less xylene is taken up and toluric acid excreted by women than by men (4). Naturally the fluctuation of the toluric acid concentration in the urine samples collected at the end of exposure is greater than that in the 24-hour urine which, however, cannot be collected under workplace conditions. Different amounts of fluid drunk and excreted also influence the toluric acid concentration in the urine.

It seems to be important that at a toluric acid concentration in the urine of as little as 3.1 g/l in a laboratory experiment a tendency towards a decrease in the systolic blood pressure was found (36).

In table 1B xylene levels in the blood which were measured in different groups of persons after differing exposure have been listed. It is surprising that after the same or supposedly the same external conditions of exposure to the toxic substance, the xylene concentration in the blood fluctuates very widely. If the blood concentrations presented in table 1B are recalculated for an ambient air concentration of 100 ml/m^3 then blood xylene concentrations between 0.09 and 5.8 mg/l result! Elimination of values over 5 mg/l, which are possibly caused by the greatly fluctuating exposure concentrations in these experiments, the great physical stress and the short exposure times, results in values mostly between 0.9 and 2.5 mg/l.

This still great range of deviation is caused by various factors. The research group of Savolainen and Riihimäki were able to show that at the same average ambient air concentration the blood xylene levels are higher the more the exposure concentration fluctuates (37, 38). To a great extent the xylene concentration in the blood is dependent on physical activity. In persons carrying out physical work Savolainen et al. (38) found that the blood xylene concentrations were more than twice those in persons at rest.

As the majority of the investigations described in the literature to date were carried out under laboratory experimental conditions, an investigation by members of the Commission on 35 painters in two factories and altogether six different workplaces is of particular importance (41). This is particularly true as the investigations made use of personal air sampling. According to these results, obtained under workplace conditions with an average concentration of the toxic substance in the ambient air at the level of the MAK value, blood xylene concentrations between 0.9 and 1.9 mg/l are to be expected.

Of great importance are the results of Savolainen who observed disturbances in the vestibular function at blood xylene concentrations of as little as 1.0 mg/l (37, 38).

4 Selection of the Indicators

For the biological monitoring of persons exposed to xylene the following parameters are available:

- determination of the xylene concentration in the alveolar air
- determination of the xylene concentration in the blood
- determination of the toluric acid excretion in the urine
- determination of the dimethylphenol excretion in the urine

Fundamentally the concentration of organic solvent in the alveolar air represents a diagnostic and sensitive parameter for objectifying the exposure to the toxic substance. Although recently there has been progress in the development of methods, this procedure cannot yet be applied as there is insufficient data for the establishment of a BAT value.

The xylene concentration in the blood represents an analytically and diagnostically satisfactory parameter for evaluating the individual xylene exposure (c.f. also BAT value toluene) and therefore seems suitable for biological monitoring especially since the toxic effect of this substance on the CNS is caused by the unchanged molecule. The necessity for blood sampling and the rapid decrease in the blood concentration of xylene after the end of exposure are the disadvantages of this method. If possible, the sampling of blood must therefore take place directly at the workplace.

The concentration of toluric acid in the urine represents, despite the great range of scatter of the results due to the differing dilution of the urine, a very suitable parameter for biological monitoring. The patient is hardly inconvenienced by the sampling of urine. Furthermore, since toluric acid is not excreted in the urine physiologically its occurrence is a diagnostically specific parameter of xylene exposure.

The determination of dimethylphenols in the urine also seems to be desirable. The suspicion that xylene can cause a disturbance in the function of the haematopoietic system has so far not been dispelled. The metabolic precursors of the xylenols are possibly closer to the active principle responsible for this disturbance than xylene itself or its main products of metabolism, the toluric acids. Due to the lack of data, at present the establishment of a biological threshold value for the dimethylphenol excretion in the urine does not seem to be possible.

5 Methodology

For the determination of toluric acids in urine above all gas chromatographic and high pressure liquid chromatographic methods are suitable. The working group "Analytische Chemie" ("Analytical Chemistry") of the DFG Commission for the Investigation of Health Hazards of Chemical Compounds in the Work Area has published in its collection "Analysen in biologischem Material" ("Analyses of Hazardous Substances in Biological Materials") a gas chromatographic method which fulfils the requirements of analytical reliability. After methylation of the free acids, the three isomeric toluric acids can be separated from one another and from hippuric acid by gas chromatography (42).

A high pressure liquid chromatographic method has also been published by the working group in its loose-leaf collection. It has the advantage that a derivatisation of the free acids can be dispensed with.

The gas chromatographic head-space technique is the preferred method for the determination of xylenes in blood. Such a procedure has been published in the loose-leaf collection "Analysen in biologischem Material" ("Analyses of Hazardous Substances in Biological Materials"). The working group "Analysen in biologischem Material" has also prepared a capillary gas chromatographic head-space method which allows the determination of most of the common organic solvents in blood in one analytical procedure. Using this procedure the isomeric xylenes can also be separated and determined.

6 Evaluation of the BAT Value

The results described in the literature (Tab. 1A) allow a relationship between the xylene concentration in the ambient air and the toluric acid concentration in the urine to be established. According to this, a metabolite excretion of

2 g toluric acids/l urine (2 g methylhippuric acids/l)

corresponds to an 8-hour xylene exposure at the level of the currently valid MAK value (100 ml/m^3).

This value is set as the BAT value.

At present there are no observations indicating the occurrence of disturbances in physical functions of toluric acid excretion at this level.

A relationship can also be calculated between the external exposure and the xylene concentration in the blood (Tab. 1B). According to this a concentration of

1.5 mg xylene/l blood

corresponds to a xylene exposure at the level of the MAK value. This value is also set as a BAT value.

Under laboratory experimental conditions Savolainen (37, 38) was able to demonstrate disturbances in the sense of balance at blood xylene concentrations of 1 mg/l. These cannot, however, be used for the setting of an occupational medical threshold limit value as even during the short term exposures to which the test persons were subjected habituation was seen.

Broadly based epidemiological investigations on the xylene exposure at the workplace are lacking. This is lamentable because these substances are of extraordinary importance for occupational and environmental medicine. The BAT value presented here must therefore be based above all on the effects of the xylenes which can be observed in laboratory experiments.

If under workplace conditions with concentrations below 100 ml/m^3 the changes in the blood count described from time to time and disturbances in the vegetative nervous system should be confirmed in future, a reduction of the BAT values might be necessary.

7 Interpretation of the Data

The urine samples for the determination of the toluric acid concentration should be collected at the end of exposure. Of course the metabolite concentration in the spontaneous urine samples is more likely to be subject to distorting influences (e.g. different amounts of fluid drunk and excreted etc.) than the 24-hour or 8-hour urinary excretion. Urine collection over these periods of time, however, cannot be carried out under workplace conditions. For practical reasons the collection of urine over a defined time is avoided due to the low probability that the collection is carried out reliably. This is also the reason against expressing the excreted amount in terms of time instead of volume. Furthermore, correction for specific weight and the creatinine content of the urine does not seem to be useful (40). However, the density of the urine should be used as an exclusion criterion in that samples with a density below 1.010 are not included in the evaluation. All these reasons lead to a wide fluctuation of the toluric acid concentrations in the spontaneous urine samples even under comparable conditions of exposure. For these reasons merely a single measurement which is numerically above or below the BAT value cannot be regarded as "adhering to" or "exceeding" it. This is only possible if the single measurement is not within the fluctuation range of the toluric acid concentration. According to the values described in the literature in all probability when the value measured is less than 0.5 or more than 4 g/l the BAT value is observed or exceeded, respectively.

Similar considerations also apply for fluctuation of the blood xylene concentration within wide limits. After a single measurement adherence to or exceeding of the BAT value can only be recognised if the xylene concentration in the blood is less than 0.4 or more than 3.0 mg/l, respectively. Due to the rapid decrease in the xylene concentration after the end of exposure blood sampling directly at the workplace at the end of exposure is recommended.

8 References

1. Šedivec, V., Flek, J.: Int. Arch. Occup. Environ. Health *37* (1976), 205–217
2. Šedivec, V., Flek, J.: Int. Arch. Occup. Environ. Health *37* (1976), 219–232
3. Åstrand, I., Engström, J., Övrum, P.: Scand. J. Work Environ. Health *4* (1978), 185–195
4. Senczuk, W., Orlowski, J.: Brit. J. Ind. Med. *35* (1978), 50–55
5. Riihimäki, V.: Scand. J. Work Environ. Health *5* (1979), 135–142
6. Engström, K., Husman, K., Riihimäki, V.: Int. Arch. Occup. Environ. Health *39* (1977), 181–189
7. Riihimäki, V., Pfäffli, P.: Scand. J. Work. Environ. Health *4* (1978), 73–85
8. Angerer, J.: Schriftenreihe Arbeitsmed. Sozialmed. Präventivmed. Bd. 71. Gentner Verlag, Stuttgart 1983
9. Riihimäki, V., Pfäffli, P., Savolainen, K., Pekari, K.: Scand. J. Work Environ. Health *5* (1979), 217–231
10. Hagemann, J., Angerer, J.: Verhandlungen der Deutschen Gesellschaft für Arbeitsmedizin e.V., 18. Jahrestagung in Frankfurt/M.-Hoechst vom 24.–27. 5. 1978. Loskant, H. (ed.). Gentner Verlag, Stuttgart 1978, pp. 365–371
11. National Institute for Occupational Safety and Health (NIOSH): Criteria for a recommended standard occupational exposure to xylene. H. E. W. Publication; No. 75/168 (1975)
12. World Health Oraganization (WHO): Study group on internationally recommended health based occupational exposure limits to selected organic solvents. Geneva 17.–23. 6. 1979
13. Henschler, D. (ed.): Xylol (alle Isomeren). Gesundheitsschädliche Arbeitsstoffe. Toxikologisch-arbeitsmedizinische Begründungen von MAK-Werten. 9th issue. VCH Verlagsgesellschaft, Weinheim 1983
14. Deutsche Gesellschaft für Mineralölwissenschaft und Kohlechemie: Wirkung von Xylol auf Mensch und Tier. Forschungsbericht, (1984), 174-8
15. Koelsch, F.: Handbuch der Berufskrankheiten, 2nd ed. Fischer Verlag, Jena 1959
16. Browning, E.: Toxicity and metabolism of industrial solvents. Elsevier, Amsterdam 1965, pp. 77-79
17. Sandmeyer, E. E. in: Patty's Industrial Hygiene and Toxicology, Vol. II B. Wiley & Sons, New York 1981
18. Valentin, H.: ASP *5* (1979), 105–110
19. Carpenter, C. P., Kinkead, E. R., Geary, D. L., Sullivan, L. J., King, J. M.: Toxicol. Appl. Pharmacol. *33* (1975), 543–558
20. Nelson, K. W., Ege, F. J., Ross. U., Woodman, L. E., Silverman, L.: J. Ind. Hyg. Toxicol. *25* (1943), 282
21. Sukhanova, V. A., Makar'eva, L. M., Boiko, V. I.: Gig. Sanit. *34* (1969), 130–132
22. Boiko, V. I.: Gig. Tr. Prof. Zabol. *6* (1970), 23–25
23. Stolbova, A. I.: Sborn. Trud. Kuybishevskoyo NII Gigieny, M. *7* (1972), 200
24. Gamberale, F., Annwall, G., Hultengren, M.: Scand. J. Work Environ Health *4* (1978), 204–211
25. Moeschlin, S.: Klinik und Therapie der Vergiftungen, 1st ed., Thieme Verlag, Stuttgart 1952
26. Smolik, R., Lange, A., Zatonski, W., Juzwiak, I., Andreasik, Z.: Polski Tygodnik Lekarski *28* (1973), 769–771

27. Lange, A., Smolik, R., Zatonski, W., Szymanski, J.: Int. Arch. Arbeitsmed. *31* (1973), 37–44
28. Lange, A., Smolik, R., Zatonski, W., Glazmann, H.: Int. Arch. Arbeitsmed. *31* (1973), 45–50
29. Ungvary, G., Tatrai, E., Hudak, A., Barcza, G., Lorincz, M.: Toxicology *18* (1980), 61–74
30. Ungvary, G., Varga, B., Horvath, E., Tatrai, E., Folly, G.: Toxicology *19* (1981), 263–268
31. Krotov, I. A., Chebotar, N. A.: Gig. Tr. Prof. Zabol. *16* (1972), 40–43, cited acc. NIOSH (1975) (11)
32. Hudak, A., Ungvary, G.: Toxicology *11* (1978), 55–63
33. Hofmann, A., Norpoth, K., Bolt, H. M., Gelbke, H.-P.: ASP *18* (1983), 181–185
34. Angerer, J., Lehnert, G.: Berufliche Belastung durch Xylol und Ethylbenzol. In: Konietzko/Schuckmann (eds.), 24. Jahrestagung der DGAUM vom 2.–5. Mai 1984, Mainz, Gentner Verlag Stuttgart, S. 267–270 (1984)
35. Engström, K., Husman, K., Pfäffli, P., Riihimäki, V.: Scand. J. Work Environ. Health *4* (1978), 114–121
36. Ogata, M., Tomokuni, K., Takatsuka, Y.: Brit. J. Ind. Med. *27* (1970), 43–50
37. Savolainen, K., Riihimäki, V., Seppäläinen, A. M., Linnoila, M.: Int. Arch. Occup. Environ. Health *45* (1980), 105–121
38. Savolainen, K., Riihimäki, V., Linnoila, M.: Int. Arch. Occup. Environ. Health *44* (1979), 201–211
39. Ogata, M., Yamazaki, Y., Sugihara R., Shimada, Y., Meguro, T.: Int. Arch. Occup. Environ. Health *46* (1980), 127–139
40. Norpoth, K., Heger, M.: Kreatinin als Bezugsgröße bei der Angabe von Stoffkonzentrationen im Harn. In: Henschler, D., Lehnert, G. (eds.): Biologische Arbeitsstoff-Toleranz-Werte und Expositionsäquivalente für krebserzeugende Arbeitsstoffe (EKA). Arbeitsmedizinisch-toxikologische Begründungen, 2nd issue. Deutsche Forschungsgemeinschaft, VCH Verlagsgesellschaft, Weinheim 1984
41. Angerer, J., Lehnert, G.: Verhandlungen der Deutschen Gesellschaft für Arbeitsmedizin e.V., 24. Jahrestagung in Mainz vom 2. bis 5. Mai 1984. Konietzko, J. (ed.). Gentner Verlag, Stuttgart 1984
42. Angerer, J., Schaller, K. H., Wardenbach P.: *o-*, *m-*, *p*-Tolursäure, Hippursäure in Harn. In: Henschler, D. (ed.): Analytische Methoden zur Prüfung gesundheitsschädlicher Arbeitsstoffe. Bd. 2: Analysen in biologischem Material, 6th issue. Deutsche Forschungsgemeinschaft, VCH Verlagsgesellschaft, Weinheim 1982

Author: J. Angerer
Approved by the Working Group: 17th February 1984

Tab. 1A: Investigations on xylene exposure of humans – A: Toluric acid excretion in urine

Collective; Duration of exposure; Substance	Xylene concentration air [ml/m^3]	Toluric acid	Sampling time	Comments	Literature
23 ♂ test persons; 7 h; *m*-xylene					Ogata, Tomokuni, Takatsuka [1970] (36)
a)	100	3140 mg/l	during the last 4 h of exposure	Decrease in the systolic blood pressure	
b)	200	5790 mg/l	during the last 4 h of exposure	Increase in the reaction time	
4 ♂ test persons; 8 h; *o*-xylene, *m*-xylene, *p*-xylene	50	929.5 mg/24 h	during and 20 h after exposure	Respiratory minute volume 9 l/min	Šedivec, Fleck [1976] (2)
		527.4–2259.1 mg/l	during the last two hours of exposure	Respiratory minute volume 9 l/min	
6 ♂ test persons; 8 h; *o*-, *m*-, *p*-xylene	75	approx. 1130 mg/ 24 h	during and 18 h after exposure	at rest	Šedivec, Fleck [1976] (2)
15 painters (14 ♂; 1 ♀) 8 h; *o*-xylene, *m*-xylene, *p*-xylene, ethyl benzene	50	665 mg/g creatinine	during the last 3 h of exposure	The concentration in the urine was probably measured 15 % too low	Engström, Husman, Pfäffli, Riihimäki [1978] (35)
10 test persons (♂; 5 ♀) 8 h; *m*-xylene				Respiratory minute volume 7.6 l/min ♀ 9.6 l/min ♂	Senczuk, Orlowski [1978] (4)
a)	22.7	♂ 532.5 mg/24 h ♀ 429.3 mg/24 h	during and 16 h after exposure		
b)	68.1	♂ 1489.3 mg/24 h ♀ 1137.3 mg/24 h			
c)	136.2	♂ 2838.4 mg/24 h ♀ 2111.9 mg/24 h			

Tab. 1A: Continued

Collective; Duration of exposure; Substance	Xylene concentration air [ml/m^3]	Toluric acid	Sampling time	Comments	Literature
18 ♂ test persons; 6 h/d; 5 d; *m*-xylene					Riihimäki, Pfäffli Savolainen [1979] (9)
a)	90; constant	1427 mg/24 h	during and 18 h after exposure	Periodic physical work	
b)	93; 70–200	1610 mg/24 h		Periodic physical work	
c)	94; constant	1152 mg/24 h		(at rest)	
4 ♂ test persons; 3 h; *m*-xylene	138	1639 mg/l 479 mg/24 h	end of exposure		Ogata, Yanazaki, Sugihara, Shimada, Meguro [1980] (39)
4 ♂ med. tech. assistants; 8 h; *o*-xylene, *m*-xylene, *p*-xylene, ethyl benzene					Angerer [1983] (8)
a) 3 ♀	81	337–1201 mg/l	during the last 1.5 h of exposure		
b) 1 ♀	66	517 mg/l			
35 ♂ painters; 8 h; *o*-xylene, *m*-xylene, *p*-xylene ethyl benzene;	4.6–19.4	286–588 mg/l	end of exposure	The concentrations stated are average values at the individual work-places	Angerer [1984] (34)
6 workplaces	11.9	383		Average of all workplaces	

Tab. 1B: Investigations on xylene exposure of humans – B: Level of the toxic substance in the blood

Collective; Duration of exposure; Substance	Xylene concentration air [ml/m^3]	Xylene concentration in the blood [mg/l]	Sampling time	Comments	Literature
15 painters (14 ♂ ; 1 ♀) 8 h; *o*-xylene, *m*-xylene, *p*-xylene			1 min after exposure	Authors register great fluctuations in the blood concentration	Engström, Husman, Pfäffli, Riihimäki [1978] (35)
	1–20	0.07–0.42			
	21–40	0.33–0.62			
	41–60	0.31–1.36			
12 ♂ test persons; 2 h; *o*-xylene, *m*-xylene, *p*-xylene, ethyl benzene			at the end of exposure	90 min physical work 30 min rest	Åstrand, Engström, Övrum [1978] (3)
a)	196	6.1			
b)	98	5.2			
6 ♂ test persons; 6 h/d; 8 d; *m*-xylene			at the end of exposure	at rest	Savolainen, Riihimäki, Linnoila [1979] (38)
a)	100	1.05			
b)	200	2.11		Disturbances in the vestibular function	
c)	200 (130–400)	3.19		Severe disturbances in the sense of balance Feeling of drunkenness Reduced reaction time	
18 ♂ test persons; 6 h/d; 5 d; *m*-xylene			at the end of exposure		Riihimäki, Pfäffli, Savolainen [1979] (9)
a)	90	up to 2.01		Periodic physical work	
b)	93 (70–200)	up to 5.40		at rest	
c)	94	up to 1.06			

Tab. 1B: Continued

Collective; Duration of exposure; Substance	Xylene concentration air [ml/m^3]	Xylene concentration in the blood [mg/l]	Sampling time	Comments	Literature
8 ♂ test persons; 6 h/d; 6 d; *m*-xylene			continuously during exposure	Physical work: 100 W; 4 times a day 10 mins	Savolainen, Riihimäki, Seppäläinen, Linnoila [1980] (37)
a)	90	x = 1.7		Disturbances in the sense of balance, increase in the reaction time, disturbance in manual co-ordination	
b)	93 (64–200)	x = 2.3		Disturbance in the vestibular function, increase in the reaction time	
		as from 1.0		Disturbance in the vestibular function	Savolainen quoted in Savolainen et al. [1980] (37)
35 ♂ painters; 8 h/d; *o*-xylene, *m*-xylene, *p*-xylene, ethyl benzene; 6 workplaces			at the end of exposure		Angerer [1984] (34)
	4.6–19.4	0.058–0.226		Single values evaluated at the individual workplaces	
	11.9	0.164		Average of all persons and workplaces	

Documentation for Carcinogenic Substances

With Biological Exposure Equivalents

Arsenic trioxide

EKA The correlation between external and internal exposure yields the following data:

Air Arsenic trioxide mg As/m^3	Urine Arsenic µg As/l
0.01	50
0.05	90
0.10	130

Sampling: end of exposure or end of shift after several consecutive shifts

Date of evaluation 1993

Synonyms –

Formula As_2O_3

Molecular weight 197.84

Melting point
amorphic or glassy form 312.3 °C
crystalline cubic form 274.0 °C

Solubility in water
amorphic or glassy form 37 g/l (20 °C)
crystalline cubic form 18 g/l (20 °C)

MAK [last established: 1975] Carcinogenic substance: Category III A1 of the MAK and BAT Values List

TRK [1993] 0.1 mg/m^3 (calculated as arsenic in total dust) for arsenic trioxide and pentoxide, arsenous acids, arsenic acid and its salts (arsenites, arsenates)

In 1986 the working group "Setting of Threshold Limit Values in Biological Material" of the DFG Commission for the Investigation of Health Hazards of Chemical Compounds in the Work Area published a relationship between the arsenic concentration in the air and the level of arsenic excretion in biological material for exposure to arsenic trioxide at the workplace (Exposure Equivalents for Carcinogenic Working Materials, EKA). The volatile arsenic compounds in the urine sample which can be determined after direct hydrogenation were defined as the arsenic excretion in urine. These are inorganic arsenic and its products of metabolism monomethylarsonic acid and dimethylarsinic acid. The correlation determined between the external and internal exposure is based on the data from two field studies (31, 36).

The TRK value valid at the time of the evaluation was 0.2 mg/m^3. Not least thanks to a publication by the working group "Setting of Threshold Limit Values in Biological Material" the TRK value was reduced in 1988 to 0.1 mg/m^3 (39).

In the meantime the availability of new scientific data allows reassessment of the relationship between external and internal exposure to arsenic. Of particular relevance is a study on the external and internal exposure to arsenic trioxide by Offergelt et al. (30). Arsenic concentrations in the air are shown to correspond to lower renal excretion levels than described in the relationship between exposure equivalents. The reasons for this difference include the fact that the method of determination of arsenic in the air at the workplace used to ascertain the external exposure has been improved. A method published by the ISO allows the efficient determination of arsenic trioxide dusts and fumes in the air. This is achieved by the use of specifically prepared filters. It can be assumed that in most previous investigations the published arsenic concentrations in the air were too low with the consequence that the arsenic excretion levels given in the correlation tables should be assigned to higher air concentrations.

This new data from the literature and the improvements in the methods make a re-evaluation of the EKA values for arsenic excretion in urine necessary.

1 Metabolism and Kinetics

Arsenic compounds can be taken up by man both via the lungs and also via the gastrointestinal tract. At the workplace inhalation is the dominant route of absorption. It should be borne in mind that in addition to inhalation, gastrointestinal absorption of arsenic from contaminated hands must also be assumed to take place (36).

The pulmonary deposition of inhaled arsenic compounds is above all determined by the size of the particles (32, 34). Non-inspirable particles are transferred with the mucous into the gastro-intestinal canal. From there, transfer of arsenic trioxide or the arsenic acid via the intestinal mucous membranes to the blood and lymph vessels is possible (34). The absorption in the lungs depends to a great extent upon the solubility of the arsenic compounds. Foa et al. (18) have suggested that pulmonary clearance is biphasic. 75 % is eliminated with a half-time of four days, 25 % with a half-time of ten days. Studies on persons who were investigated 2 to 19 years after the end of exposure provided indications of considerably longer retention times (8).

Gastrointestinal absorption is the most important route of absorption of arsenic from non-occupational exposure. Absorption takes place above all with food, in particular when sea-food is eaten, and with drinking water. The absorption of inorganic arsenic compounds from the gastrointestinal tract is determined mainly by the solubility of the compounds. The proportion of the dose absorbed in humans is more than 90 % if water-soluble compounds are ingested. In the case of arsenic trioxide which is only slightly soluble in water the gastro-intestinal absorption is slower and depends on the size of the particles and the pH of the hydrochloric acid in the stomach. Organic arsenic compounds from sea-food are almost completely absorbed after ingestion (18).

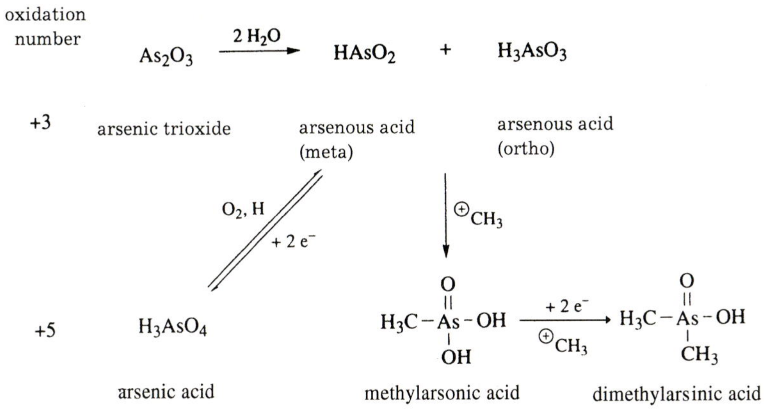

Fig. 1: Metabolism of arsenic trioxide in mammals and humans; metabolites in the urine: arsenic (III), arsenic (V), methylarsonic acid (MA) and dimethylarsinic acid (DMA)

The metabolism of arsenic in the human organism is shown in figure 1. Animal experimental and human investigations have shown that arsenic (III) is metabolised to arsenic (V), monomethylarsonic acid (MMA) and dimethylarsinic acid (DMA). The main metabolite is dimethylarsinic acid which accounts for 50–80 % of the urinary metabolites. It is in principle true that occupationally exposed persons show the same spectrum of metabolites as non-exposed persons. However, there are quantitative differences which depend on the level of exposure. After high exposures the metabolism to dimethylarsinic acid decreases which is probably to be interpreted as a saturation of the methylation process (5, 6, 15, 24). Methylation is the most important natural detoxification mechanism. Both monomethylarsonic acid and dimethylarsinic acid are less toxic than the inorganic acid.

Organoarsenic compounds such as occur in food are not biologically transformed and thus form no inorganic or methylated arsenic compounds in the human organism.

Buchet et al. (4) investigated the changes with time in the concentration of arsenic metabolites in the urine after ingestion (3 persons) of 6.67 μmol sodium arsenate (III). The total arsenic content was determined by dry ashing, the metabolites arsenous acids and arsenic acid as well as monomethylarsonic acid (MMA) and dimethylarsinic acid (DMA) according to a modified method from Braman et al. (7). After one day, 22 % of the dose had been excreted, 38 % in the form of inorganic arsenic, 23 % as MMA and 39 % as DMA. After four days, 50 % of the arsenic taken up had been excreted with the urine, whereby the proportion of inorganic arsenic was reduced to 11 % and that of dimethylarsinic acid had increased to 76 %. After 14 days the persons showed normal arsenic values.

Mappes (26) obtained the following results after ingestion of 13.40 μmol arsenic trioxide in an experiment on himself: 52 % of the arsenic taken up had been excreted in the urine after 24 hours. Pharmacokinetic investigations demonstrated that for arsenic trioxide there are two different compartments.

2 Critical Toxicity

Chronic arsenic intoxication is of importance from an occupational medical point of view. A detailed description of the symptoms can be found in the occupational-medical toxicological documentation of the MAK value (19) as well as in Landrigan (22), Arnold (3) and Vahter (35).

Due to its carcinogenic properties in humans, arsenic trioxide as well as arsenic pentoxide, arsenous acid, arsenic acid and its salts have been classed in category A1 of section III "Carcinogenic Substances" of the MAK and BAT Values List (13). A review of the human carcinogenicity can be found in the documentation of the IARC (21) and the WHO (38).

3 Exposure and Effects

The DFG Commission for the Investigation of Health Hazards of Chemical Compounds in the Work Area have given their detailed opinion on the question of the body burden and effect on persons occupationally exposed to arsenic trioxide both in the occupational-medical toxicological documentation of MAK Values and also in the publication by Wichmann and Lehnert (39). Due to the carcinogenic effects of inorganic arsenic in man the chronic effects of arsenic cannot be used to evaluate a threshold limit value.

Enterline found in an epidemiological study a significant linear relationship between the standardised mortality ratio (SMR) for lung cancer and the renal arsenic excretion (14). On the basis of the regression equation of SMR = 86.1 + 0.590 As in urine (µg/l) the SMR for an arsenic concentration in the urine of 100 µg/l is 145, for 500 µg/l the SMR is 380.

The US American BEI documentation (2) is based on the relationship between internal exposure, determined as arsenic excretion in the urine, and the corresponding effects on health. As relevant effects peripheral neuropathy, peripheral angiopathy and liver and kidney function disorders are listed.

4 Selection of the Indicators

After exposure to inorganic arsenic there is a significant and specific increase in the arsenic species monomethylarsonic acid, dimethylarsinic acid and inorganic arsenic excreted in the urine. The *determination* of these three arsenic species *in the urine* is therefore the preferred method for the biological monitoring of workers who are exposed to inorganic arsenic. The measurement of these species of arsenic is not influenced by organic arsenic compounds taken up with food.

Inorganic arsenic and its metabolites are rapidly excreted after the beginning of exposure. The urine concentration increases slowly and remains at a relatively constant level during the first days of exposure. During the working day and from the end of work to the beginning of the next shift there are no great changes in the concentration (30, 36). The elimination kinetics result in a significant accumulation of arsenic and its metabolites during the working week. Sampling should therefore take place at the end of the working week.

The determination of *arsenic in blood* is of no importance in the biological monitoring of occupationally exposed persons (22, 28). The reason for this is that the biological half-life in blood both for inorganic and organic arsenic is relatively short. This means that the blood concentration is only at an increased level for a short time after absorption. Furthermore, there are no routine methods available for the specific determination of inorganic arsenic and its metabolites in blood.

The arsenic concentrations in *hair and finger nail samples* are usually higher than in other compartments of the human organism. The reason for this is the high content of cystine whose thiol groups bind trivalent inorganic arsenic. The measurement of the arsenic content of hair and finger nail samples has not acquired any importance in occupational medical practice, the reasons being the non-standardised sampling, difficulties in analysis as well as a considerable danger of contamination. Moreover, there are no relevant studies for the evaluation of threshold limit values.

5 Methodology

Inorganic and organic arsenic compounds in urine are transformed for quantitative measurement into volatile hydrides and determined by atomic absorption spectroscopy (hydride method) (16, 17). Using the method of direct hydrogenation of a urine sample only inorganic arsenic compounds and their metabolites monomethylarsonic acid and dimethylarsinic acid are transformed into volatile hydrides and quantified. Organic arsenic compounds, which are taken up during consumption of sea-foods and excreted renally, are not determined (29, 37). A tested method can be found in the loose-leaf collection “Analysen im biologischen Material” (20) and in its English edition “Analyses of Hazardous Substances in Biological Materials” (1).

If the urine sample is digested wet or dry the total arsenic content is determined. Therefore wet or dry digestion should not be carried out.

The following explanations refer to the measurement of inorganic arsenic and its metabolites. In biological monitoring this “arsenic fraction” should be measured. Table 1 gives an overview of the published data for non-exposed persons. Only studies which distinguish between the various arsenic species were taken into account. Column Σ shows the “fraction” determined by the hydride AAS method.

Tab. 1: Arsenic values in the air and urine in non-exposed persons and percentage distribution of the various arsenic species

Air	Urine							Literature
($\mu g\ As/m^3$)	As (III) (µg/l)	As (V) (µg/l)	Methylarsonic acid (MA) (µg/l)	Dimethylarsinic acid (DMA) (µg/l)	Σ As (III), As (V), MA, DMA (µg/l)	Total As (µg/l)	spec. weight urine	
	8.4 % 1.9	17.3 % 3.9	8.0 % 1.8	66.4 % 15.0	100 % 22.6	–	–	4[1)]
			3.6 ± 2.4	15.5 ± 6.8				4[2)]
	6 % 3.7 ± 1.4 6 %	5 % 3.0 ± 1.3 5 %	20 % 11.3 ± 5.3 19 %	68 % 39.2 ± 18.2 67 %	100 % 57.3 ± 21.3 98 %	58.4 100 %	1.024	48[3)]
3.6 ± 1.56[a)]	7 % 1.3 ± 1.58 6 %	7 % 1.3 ± 1.59 6 %	19 % 3.4 ± 1.63 16 %	66 % 11.5 ± 1.47 54 %	100 % 17.5[b)] 83 %	21.2 100 %	1.018	38[4)]
	11 % 1.6 ± 1.2 11 %		12 % 1.7 ± 1.2 11 %	77 % 11.2 ± 5.9 74 %	100 % 14.5 ± 7.5[a)] 96 %	15.1[b)] 100 %	–	6[5)]

[1)] n = 4.3 ♂ + 1 ♀, 34 ± 7 8 years
[2)] x ± s, n = no data
[3)] Ibd. p. 51, Tab. 2, x ± s, n = 20; total arsenic was calculated on the basis of a given correlation: $y_{wet\ digestion}$ (µg/l) = 0.915 · $x_{separation\ method}$ (µg/l) + 6
[4)] a) geom. mean ± s, 56 % <1.2 $\mu g/m^3$ (detection limit), n = 41; urine values: geom. mean ± s, n = 41; b) calculated; total arsenic determined by wet digestion
[5)] Ibd. p. 22, Tab. 10, x ± s, n = 5; a) calculated b) dry ashing; geom. mean because an outlier is included (arithm. mean 18.7 ± 14.3)

As the measurement of arsenic in the air is also of great importance for the evaluation of a correlation between the external and internal exposure the method is briefly described here. The determination of arsenic in particle and vapour form should take place according to the sampling method AGSA-AA I No. 15-90 of the ISO. Cellulose acetate membrane filters impregnated with sodium carbonate and glycerine are used as sampling phases. The filters prepared in this way guarantee an extremely efficient separation of the arsenic compounds, independent of the state of aggregation (10, 30). For the quantitative determination of arsenic the filters are digested by wet oxidative digestion and the arsenic quantified using hydride AAS or GF-AAS. This method represents the present state of the art.

6 Background Exposures

Arsenic can be taken up by man in drinking water, food and from the air. Food is the main route of absorption. Food, with the exception of sea-food, contains less than 0.25 mg arsenic/kg; various kinds of fish between 1 and 10 mg/kg and shell fish more than 100 mg/kg. In food of maritime origin the arsenic is in organic form (arsenobetaines and arsenocholines). The daily uptake via food is estimated to be between 0.04 (without fish) and 0.19 mg arsenic (with fish). The daily arsenic uptake via the drinking water and the air is not significant.

With the hydride AAS method suggested in chapter 5 the aromatic arsenic compounds excreted unmetabolised in the urine are not determined. Under these conditions of measurement the "normal" arsenic excretion is below 25 µg/l (see also Tab. 1). Thus, it must be ensured that the above-mentioned method of analysis is used for the monitoring of persons occupationally exposed to arsenic. After the digestion of urine samples and the consumption of crustaceans arsenic can be measured in the urine in the mg range.

7 Evaluation of the Biological Exposure Equivalents for Carcinogenic Substances (EKA)

Various field studies on humans can be referred to for the evaluation of correlations between the external arsenic exposure and the renal excretion of arsenic. The correlation evaluated in 1986 is based mainly on two studies (31, 36).

The meaningfulness of the field study carried out by Pinto et al. (31) on 24 workers from a smelting works is limited by the fact that the total arsenic excretion in the urine was analysed without any further specification. It was checked anamnestically that no fish was eaten in the period of investigation. Nevertheless, it cannot be excluded that the results were influenced by the uptake of organoarsenic compounds. Furthermore, it must be assumed that the arsenic concentrations given for the air were too low as arsenic oxide vapours were probably not determined.

This assumption also applies to the study by Vahter et al. (36). According to investigations by Costello et al. (10) only approx. 42 % of the arsenic trioxide in the vapour phase is determined when untreated cellulose acetate filters are used. There was a highly significant correlation of r = 0.92 between the arsenic concentration in the air measured by personal air sampling and the renal excretion of organic arsenic and its metabolites (36). However, with four persons the urinary excretion values were much greater than were to be expected from the external exposure. The authors suspected that the persons had been exposed to arsenic taken up directly from contaminated hands, cigarettes or snuff. Therefore this study was not included in the present evaluation.

The study carried out by Landrigan et al. (23) was also not included in the evaluation. This working group investigated 48 lead battery workers who were exposed both to arsine and arsenic trioxide. According to the currently available data it seems that the relationship between the arsenic content of the air and the renal excretion of inorganic arsenic and its metabolites is influenced by the chemical form of the arsenic species at the workplace. It is known e.g. that lead arsenate, calcium arsenate, gallium arsenide and to a certain extent arsenic trisulphide are absorbed more slowly in the lungs than arsenic trioxide. Arsine, however, is absorbed considerably better than arsenic trioxide. Due to this mixed exposure and the fact that the total arsenic content of the urine was determined, this study can also not be used for the evaluation.

Roels at al. (33) investigated persons exposed to arsenic in a glass factory. As, on the one hand, only a small fraction of the dust was able to enter the alveoli and, on the other, the urinary excretion levels were relatively high, the authors assumed that oral uptake from contaminated hands had had a considerable influence on the results.

The studies described below were taken into account for the evaluation of the EKA value. The data presented there are used for the compilation of table 2.

Tab. 2: Calculated arsenic concentration in the urine (arsenic, inorganic and metabolites) at various arsenic concentrations in the air at the workplace

	Arsenic concentrations in the air (mg/m^3)			
Study	0.01	0.05	0.10	Reference
82 smelters As_2O_3 exposure	25 µg/l	76 µg/l	139 µg/l	Smith 1977
28 smelters As_2O_3 exposure	44 µg/l	101 µg/l	127 µg/l	Enterline 1987
11 smelters As_2O_3 exposure	68 µg/l	102 µg/l	146 µg/l	Yamauchi 1989
22 employees of sulphuric acid production Exposure to As_2O_3 dust and fumes	30 µg/g creatinine	54 µg/g creatinine	69 µg/g creatinine	Offergelt 1992

In a comprehensive study Smith et al. (34) investigated the excretion of inorganic arsenic and its metabolites in the urine of 82 smelters employed in a copper works. The external exposure was determined by personal air sampling, urine sampling took place in the morning on two consecutive days. There was a statistically significant correlation between the external exposure and the concentration of various arsenic species in the urine. A regression equation was only given for the relationship between the arsenic in the air and the dimethylarsinic acid excretion in urine. As dimethylarsinic acid represents approx. 65 % of the arsenic metabolites and the ratio of the arsenic species to one another was relatively constant and independent of the exposure concentration, for the fraction inorganic arsenic and its metabolites in the urine a concentration of about 70 µg/l can be estimated for external exposure to 50 $\mu g/m^3$. In addition the authors quoted values for external and internal exposure for four exposure groups:

	Arsenic in air	Arsenic in urine
Controls	3.6 $\mu g/m^3$	17.5 $\mu g/m^3$
Group with low-level exposure	8.3 $\mu g/m^3$	25.7 $\mu g/m^3$
Group with medium-level exposure	46.1 $\mu g/m^3$	49.6 $\mu g/m^3$
Group with high-level exposure	52.7 $\mu g/m^3$	96.6 $\mu g/m^3$

The data yields the following relationship as a linear regression:

Arsenic in urine (µg/l) = 12.6 + 1.26 arsenic in air ($\mu g/m^3$)

Within the framework of an epidemiological study of the frequency of lung cancer in the workers of a copper smelting works, a relationship between urine excretion and external exposure was established. The relationship was not linear: arsenic in urine [µg/l] = (arsenic in air [$\mu g/m^3$]/0.0064)$^{0.515}$. The measurements were based on stationary air sampling (14).

Yamauchi et al. (40) published in 1989 a field study on 37 gallium arsenide workers and 11 smelters. The air concentrations measured by stationary air sampling were associated with urine values which resulted from two urine collections on 2 to 3 days. For gallium arsenide there were only small increases in the arsenic excretion. This can be attributed to the fact that gallium arsenide can only be absorbed by the human organism to a very small extent (35, 40). For the smelters who were exposed to arsenic trioxide an average urine excretion of inorganic arsenic and its metabolites of 239 µg/l was associated with an average air concentration of 0.336 mg/m^3.

In a field study that was published in 1992 by Offergelt et al. (30) 18 workers of a company producing sulphuric acid were investigated. The arsenic concentration in the breathing zone was determined by the sampling and analysis method described above. The air measurements took place during five days, the urine samples were collected after the shift and before the next shift. In biological material the inorganic arsenic as well as the monomethylarsonic acid and the dimethylarsinic acid were measured selectively. For 82 pairs of values there was a significant correlation of $r = 0.48$. The regression equation was log arsenic in urine (µg/g creat.) = 1.130 + 0.353 log arsenic in

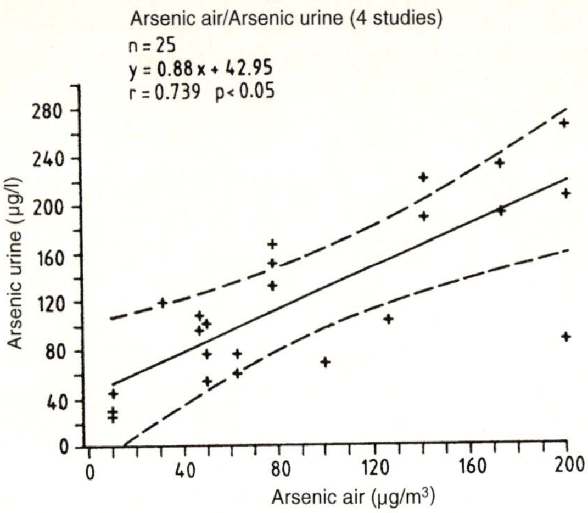

Fig. 2: Correlation diagram for the relationship between the arsenic concentration in the air and the renal arsenic excretion on the basis of the four studies used for the evaluation of the EKA values

air (µg/m^3). Table 3 shows the arsenic excretion (inorganic arsenic + metabolites) in the urine extrapolated for various arsenic concentrations in the air. The post shift values as well as the values before the next shift were taken into consideration.

The arsenic excretion levels in the urine listed in table 2 for defined arsenic concentrations in the air are calculated from linear and non-linear relationships between the data for the external and internal exposure. With the exception of the study by Enterline et al. (14), in which the arsenic fraction measured was not described, the urine excretion levels are based on the measurement of inorganic arsenic and its metabolites.

The correlates calculated show a high variance. This wide range can be explained, on the one hand, by the different methods of sampling and analysis of arsenic in the air. On the other hand the extrapolation from low external exposures to higher ones is critical with this database. As a rule the external exposures in the studies listed are below 0.05 mg/m^3. In the lower exposure range (0.01–0.05 mg/m^3) the calculated values correspond better than those in the upper range.

The lowest excretion values are to be found in the study by Offergelt et al. (30) in which great importance was laid on an exact and complete determination of the arsenic in the air. For a period of five days the external exposure was determined with a method of sampling and analysis recommended by the ISO, with personal air sampling during the whole of the shift. Of the persons investigated no-one wore personal breathing protection. An oral uptake of arsenic can be largely excluded. The arsenic levels in the urine measured at the end of the shift do not differ greatly from those in the urine samples which were collected before the next shift (see Tab. 3). The individual range of variation of the arsenic excretion is, however, considerable. It can be seen in table 3. The largest values measured in the lower range of exposure correspond in their order of magnitude with the extrapolated values from the other studies.

Tab. 3: Calculated renal arsenic excretion (inorganic arsenic and its metabolites MA + DMA) in the urine samples collected post-shift and before the next shift in relation to the arsenic concentration in the air at the workplace with exposure to As_2O_3 dusts and fumes (30)

Arsenic in air (mg/m³)	Renal arsenic excretion (µg/g creatinine)	
	after the shift	before the next shift
0.0125	33 (26–42)	31 (24–40)
0.025	42 (36–50)	39 (33–47)
0.050	54 (47–62)	50 (43–58)
0.100	69 (56–84)	64 (52–79)
0.200	88 (65–117)	81 (60–111)
0.400	112 (75–166)	104 (68–158)

The results are given as geometric means with 95 % confidence intervals

The evaluation of the EKA values takes place on the basis of the four selected studies. The correlation analysis yields the relationship between air and urine values shown in figure 2. Only the data up to an external exposure to 0.20 mg/m³ are included. The arsenic excretion in the urine (inorganic arsenic and metabolites) was determined by interpolation for external exposures of 0.01, 0.05 and 0.1 mg/m³. The relationship between the arsenic concentration in the air at the workplace and the arsenic excretion in the urine is shown below:

Air Arsenic concentration	Urine Arsenic concentration
0.01 mg/m³	51 µg/l
0.05 mg/m³	87 µg/l
0.10 mg/m³	131 µg/l

With regard to the criticism of the studies used, discussed in detail above, this correlation should be regarded as provisional.

8 Interpretation of the Data

Pretreatment of urine samples by wet or dry ashing techniques permits determination of the total arsenic content of the urine. In this way the arsenic taken up with food is included. In particular fish and other sea-foods contain high levels of organic arsenic compounds (9, 25, 27) which are quickly absorbed and excreted via the kidneys (11, 12). This can lead to a great increase in the total arsenic content of the urine (4, 17, 25, 37). For this reason pretreatment by wet or dry ashing should not be carried out.

The method of direct hydrogenation of the urine sample should be used. Even with this procedure it was observed that after ingestion of sea-food there is an increase in the level of arsenic compounds. The increase is small and can amount to 12 µg As/l urine (29, 37).

Occupational medical health examinations on persons exposed to inorganic arsenic must be carried out according to the berufsgenossenschaftlicher Grundsatz G 16 (guideline issued by the Employers' Liability Insurance Association).

9 References

1. Angerer, J., Schaller, K. H.: Analyses of Hazardous Substances in Biological Materials, Volume 3. Deutsche Forschungsgemeinschaft, VCH Verlagsgesellschaft, Weinheim 1991
2. American Conference of Governmental Industrial Hygienists (ACGIH): Threshold Limit Values for Chemical Substances and Physical Agents and Biological Exposure Indices. ACGIH, Cincinnati 1991–1992
3. Arnold, W.: Arsen, ein allgegenwärtiges Gift in der Umwelt des Menschen. Öff. Gesundh.-Wes. *49* (1987), 240–249
4. Buchet, J. P., Lauwerys, R., Roels, H.: Comparison of several methods for the determination of arsenic compounds in water and urine. Int. Arch. Occup. Environ. Health *46* (1980), 11–29
5. Buchet, J. P., Lauwerys, R., Roels, H.: Comparison of the urinary excretion of arsenic metabolites after a single oral dose of sodium arsenite, monomethylarsonate or dimethylarsinate in man. Int. Arch. Occup. Environ. Health *48* (1981), 71–79
6. Buchet, J. P., Lauwerys, R.: Study of inorganic arsenic methylation by rat liver in vitro: Relevance for the interpretation of observations in man. Arch. Toxicol. *57* (1985), 125 –129
7. Braman, R. S., Foreback, C. C.: Methylated forms of arsenic in the environment. Science *182* (1973), 1247–1249
8. Brune, D., Nordberg, G., Webster, P. O.: Distribution of 23 elements in the kidney, liver and lungs of workers from a smeltery and refinery in North Sweden exposed to a number of elements and of a control group. Sci. Tot. Environ. *16* (1980), 13–53
9. Chapmann, A. C.: On the presence of compounds of arsenic in marine crustaceans and shell fish. Analyst *51* (1926), 548–563
10. Costello, R. J., Eller, P. M., Hull, R. D.: Measurement of Multiple Inorganic Arsenic Species. Am. Ind. Hyg. Assoc. J. *44* (1983), 21–28
11. Coulson, E. J., Remington, R. E., Lynch, K. M.: Metabolism in the rat of the naturally occurring arsenic of shrimp as compared with arsenic trioxide. J. Nutrition *10* (1935), 255–270
12. Crecelius, E. A.: Changes in the chemical speciation of arsenic following ingestion by man. Environ. Health Perspect. *19* (1977), 147–150
13. Deutsche Forschungsgemeinschaft: List of MAK and BAT Values 1993, Maximum Concentrations and Biological Tolerance Values at the Workplace. VCH Verlagsgesellschaft, Weinheim 1993
14. Enterline, P. E., Henderson, V. L., Marsh, G. M.: Exposure to arsenic and respiratory cancer. A reanalysis. Am. J. Epid. *125* (1987), 929–938

15. Farmer, J. G., Johnson, L. R.: Assessment of occupational exposure to inorganic arsenic based on urinary concentrations and speciation of arsenic. Brit. J. Ind. Med. *47* (1990), 343–348
16. Fishbein, L.: Overview of analysis of carcinogenic and/or mutagenic metals in biological and environmental samples. I. Arsenic, beryllium, cadmium, chromium and selenium. Int. J. Environ. Anal. Chem. *17* (1984), 113–170
17. Fleischer, M., Schaller, K. H. in: Welz, B. (ed.): 4. Colloquium Atomspektrometrische Spurenanalytik. Überlingen 1987, pp. 577–585
18. Foa, V., Colombi, A., Maroni, M., Barbieri, F., Franchini, I, Mutti, A., De Rosa, E., Bartolucci, G. B.: Study of kidney function of workers with chronic low level exposure to inorganic arsenic. In: Foa, V., Emmett, E. A., Maroni, M., Calombi, A. (eds.): Occupational and environmental chemical hazards. Wiley & Sons, New York 1987, pp. 362–367
19. Henschler, D. (ed.): Gesundheitsschädliche Arbeitsstoffe. Toxikologisch-arbeitsmedizinische Begründung von MAK-Werten, 18th issue. VCH Verlagsgesellschaft, Weinheim 1992
20. Henschler, D. (ed.): Analytische Methoden zur Prüfung gesundheitsschädlicher Arbeitsstoffe. Band 2: Analysen in biologischem Material, 10th issue. VCH Verlagsgesellschaft, Weinheim 1991
21. International Agency for Research on Cancer (IARC): Some Metals and Metallic compounds. In: The evaluation of the carcinogenic risk of chemicals to humans. IARC Monograph No. 23. IARC, Lyon 1980, pp. 39–141
22. Landrigan, P. J.: Arsenic – State of the Art. Am. J. Ind. Med. *2* (1981), 5–14
23. Landrigan, P. J., Costello, R. J., Stringer, W. T.: Occupational exposure to arsine. An epidemiologic reappraisal of current standards. Scand. J. Work Environ. Health *8* (1982), 169–177
24. Lovell, M. A., Farmer, J. G.: Arsenic Speciation in Urine from Humans Intoxicated by Inorganic Arsenic Compounds. Human Toxicol. *4* (1985), 203–214
25. Lunde, G.: Occurrence and transformation of arsenic in the marine environment. Environ. Health Perspect. *19* (1977), 47–52
26. Mappes, R.: Versuche zur Ausscheidung von Arsen im Urin. Int. Arch. Occup. Environ. Health *40* (1977), 267–272
27. Munro, I. C.: Naturally occurring toxicants in foods and their significance. Clin. Toxicol. *9* (1976), 647–663
28. Nordberg, G. F., Pershagen, G., Lauwerys, R.: Inorganic arsenic. Toxicological and epidemiological aspects. Report to the Commission of European Communities. Department of Community Health and Environmental Medicine. Odense University, Odense 1979
29. Norin, H., Vahter, M.: A rapid method for the selective analysis of total urinary metabolites in inorganic arsenic trioxide exposure. Scan. J. Work Environ. Health *7* (1981), 38–44
30. Offergelt, J. A., Roels, H., Buchet, J. P., Boeckx, M., Lauwerys R.: Relation between airborne arsenic trioxide and urinary excretion of inorganic arsenic and its methylated metabolites. Brit. J. Ind. Med. *49* (1992), 387–393
31. Pinto, S. S., Varner, M. O., Nelson, K. W., Labbe, A. L., White, L. D.: Arsenic trioxide adsorption and excretion in industry. J. Occup. Med. *18* (1976), 677–660
32. Pinto, S. S., Enterline, P. E., Henderson, V., Varner, M. O.: Mortality experience in relation to a measured arsenic trioxide exposure. Environ. Health Perspect. *19* (1977), 127–130
33. Roels, H., Buchet, J. P., Truc, J., Croquet, F., Lauwerys, R.: The possible role of direct ingestion on the overall absorption of cadmium or arsenic in workers exposed to CdO or As_2O_3 dust. Am J. Ind. Med. *3* (1982), 53–65
34. Smith, T. J., Crecelius, E. A.: Airborne arsenic exposure and excretion of methylated arsenic compounds, Environ. Health. Perspec. *19* (1977), 89–93
35. Vahter, M.: Arsenic. In: Clarkson, T. W., Friberg, L., Nordberg, G. F., Sager, P. R. (eds.): Biological monitoring of toxic metals. Plenum Press, New York 1988, pp. 303–321
36. Vahter, M., Friberg, L., Rahnster, B., Nygren, A.: Airborne arsenic and urinary excretion of metabolites of inorganic arsenic among smelter workers. Int. Arch. Occup. Environ. Health *57* (1986), 79–91
37. Valkonene, S., Jarvisalo, J., Aitio, A.: Urinary arsenic in a Finnish population without occupational exposure to arsenic. In: Brätte/Schramel (eds): Trace Element-Analytical Chemistry in Medicine and Biology, Vol. 2, de Gruyter, Berlin 1983, pp. 611–621
38. World Health Organization (WHO): Environmental Health Criteria 18. Arsenic. IPCS International programme on chemical safety (1981)

39. Wichmann, N., Lehnert, G.: Arbeitsmedizinische Erfahrungen und TRK-Wert für Arsen, Arbeitsmed., Sozialmed., Präventivmed. *22* (1987), 18–20
40. Yamauchi, H., Takahashi, K., Mashiko, M., Yamamura, Y.: Biological monitoring of arsenic exposure of gallium arsenide- and inorganic arsenic-exposed workers by determination of inorganic arsenic and its metabolites in urine and hair. Am. Ind. Hyg. Assoc. J. *50* (1989), 606–612

Author: K. H. Schaller
Approved by the Working Group: 27th May 1993

Ethylene

EKA	The correlation between external and internal exposure yields the following data:

Air ethylene (ppm)	Erythrocytes hydroxyethylvaline (nmol/g Hb)	(µg/l blood)
25	1.8	45
50	3.6	90
100	7.2	180
150	10.9	260

Sampling: no restrictions

Date of evaluation	1993
Synonyms	ethene
Formula	C_2H_4
Molecular weight	28.05
Melting point	−169 °C
Boiling point	−104 °C
MAK [last established: 1993]	Suspected carcinogen: category III B of the List of MAK and BAT Values

For a long time ethylene was regarded toxicologically as being practically inert. It was recently discovered, however, that the substance is transformed in the organism to a minor extent into ethylene oxide. Ethylene oxide is listed in category A2 of section III of the MAK list.

The rate of ethylene metabolism is low. After exposure at a level of approx. 50 ppm there is, according to toxicokinetic model calculations, an internal exposure of the organism to ethylene oxide as occurs during exposure to 1 ppm ethylene oxide, the TRK value. Exposure to ethylene is therefore regarded as involving a carcinogenic risk.

In 1993 ethylene was classified in category B of section III of the MAK list. There is no MAK value. There is also no TRK at present. Prophylaxis should therefore be based on the existing TRK value for ethylene oxide and the quantitative aspects of the metabolism of ethylene to ethylene oxide.

1 Metabolism and Kinetics

Various investigations have shown that ethylene is transformed to ethylene oxide in an initial step of metabolism. In rats the formation of ethylene oxide from ethylene could be demonstrated directly both in vitro with liver microsomes (24) and in vivo in the expired air (12). The formation of ethylene oxide in mice (10, 25) and in humans (30, 13) was concluded indirectly from the observed hydroxyethylation of various amino acids of haemoglobin. The rate of ethylene metabolism is, however, low in the species which have been investigated. Below 80 ppm in rats and 50 ppm in humans it is, according to model calculations, directly proportional to the exposure concentration (3, 7, 26, 13). In rats the metabolism of ethylene can be saturated. The maximum metabolic rate is reached in this species at approx. 1000 ppm (1, 3).

The main routes of metabolism of the primary metabolite of ethylene, ethylene oxide, are the formation of 1,2-ethanediol, catalysed by epoxide hydrolase, and the addition reaction with glutathione, catalysed by glutathione-S-transferase.

Ethylene is also formed endogenously by animals and humans (21, 6, 22, 15, 26). The pharmacokinetically calculated endogenous production rate in humans (70 kg) is about 32 nmol/h and the mean concentration in the body calculated from this 0.44 nmol/kg body weight (13).

With concentrations up to 50 ppm, approx. 2 % of inhaled ethylene is retained in humans. Ethylene accumulates in the body only to a very small extent (7, 13).

Cysteine adducts are eliminated via the kidneys (8, 34). Endogenous exposure to ethylene oxide of 0.25 nmol/kg (13) can be calculated from the endogenous production rate of ethylene and pharmacokinetic data for ethylene oxide in humans (4). During workplace exposure to 1 ppm ethylene oxide (TRK value) the calculated average internal exposure is the same as occurs during exposure to an average ethylene concentration of 50 ppm (7).

2 Critical Toxicity

The critical toxicity of ethylene is, as explained above, considered to be determined by the expected carcinogenic effect of the metabolite ethylene oxide.

After occupational exposure to ethylene there is increased adduct formation with the N-terminal valine of haemoglobin; this has been attributed to the formation of ethylene oxide (31). The values were also considerably higher in smokers who inhale ethylene in tobacco smoke (32, 2, 20, 13). There is always a certain background level of hydroxyethylated haemoglobin adducts formed from endogenously-produced ethylene and the ethylene which is present in the ambient air. It has been estimated that the endogenous production together with an external exposure to 15 ppb ethylene leads to a concentration of ethylene oxide in the body of 0.25 nmol/kg (7, 13). The level of binding of ethylene oxide to the N-terminal valine of haemoglobin was calculated from this internal concentration. The resulting values corresponded very well with the values measured (13).

In the lymphocyte-DNA of persons not occupationally exposed to ethylene a background level of 7-(2-hydroxyethyl)-guanine was also determined (14).

After exposure of mice to ethylene, alkylation was observed of the amino acids histidine, cysteine and N-terminal valine in haemoglobin as well as of guanine (N-7) in DNA from various organs (10, 25). It was concluded from a comparison with the binding characteristics of ethylene oxide that the formation of adducts by ethylene is to be attributed to its metabolite ethylene oxide.

3 Exposure and Effects

Exposure to ethylene oxide at the level of the TRK value current in 1993 of 1 ppm corresponds to an exposure to ethylene of 50 ppm.

Exposure to ethylene represents a carcinogenic risk as ethylene is metabolised to the carcinogen ethylene oxide. However, since the maximum rate of formation of ethylene oxide is low, in animal experiments with ethylene itself no increases in tumour incidence could be detected.

The MAK documentation for ethylene contains a quantitative estimate of the possible risk of cancer associated with exposure to ethylene.

4 Selection of the Indicators

Direct determination in biological material of the ethylene oxide formed after ethylene exposure is difficult due to the small amount of the substance and the short biological half-life of ethylene. Macromolecular adducts of ethylene oxide, in particular with haemoglobin, are better suited as indicators of endogenous ethylene oxide exposure.

In humans haemoglobin adducts with histidine and the N-terminal valine have been detected after exposure to ethylene oxide (5, 33, 11, 20, 16, 17, 23, 29).

A linear relationship shown in the rat between haemoglobin adducts and the external concentration of ethylene oxide after long-term exposure (19, 28) is also assumed to apply for humans. The determination of the haemoglobin adducts has therefore been used to determine exposure equivalents (32) and suggested for the biological monitoring of persons exposed to ethylene oxide (18, 11). Such biological monitoring would also be possible for ethylene as the average exposure concentration of ethylene can be calculated on the basis of the known pharmacokinetics of ethylene and ethylene oxide (31, 13).

5 Methodology

The determination of the level of haemoglobin in the blood with N-terminal N-hydroxyethylvaline has been described in various studies by the Ehrenberg research group. After exposure to ethylene oxide, hydroxyethylation of haemoglobin on the amino acids histidine, cysteine and the N-terminal valine has been detected (9, 10, 19, 25, 27). The DFG Commission for the Investigation of Health Hazards of Chemical Compounds in the Work Area has compiled a collection of reliable analytical methods in the loose-leaf collection “Analysen in biologischem Material” (“Analyses of Hazardous Substances in Biological Materials”) (35).

6 Evaluation of the Biological Exposure Equivalents for Carcinogenic Substances (EKA)

After occupational exposure to ethylene oxide below the current TRK value (1 ppm, corresponding to approx. 2 mg/m^3) there is a dose-dependent increase of N-terminal 2-hydroxyethylvaline in the haemoglobin of the erythrocytes. The Commission has access to data from factories from the German chemical industry; the most important results are summarised in table 1 (see also EKA documentation for ethylene oxide).

Tab. 1: Relationships between the external concentration of ethylene oxide in the ambient air and the erythrocyte concentration of hydroxyethylvaline in the globin in employees with continual exposure to ethylene oxide (J. Lewalter, unpublished investigations, 1992) (36)

Persons	Smoking habits	Ethylene oxide in the air (mg/ m^3)	Blood (erythrocytes) µg hydroxyethyl-valine per l blood
20	smokers	./.	5 (1–10.7)
16	non-smokers	./.	1 (<1–3)
7	smokers/non-smokers	0.1	40 (23–57)
34	smokers/non-smokers	0.3	43 (25–60)
28	smokers/non-smokers	0.8	50 (31–74)
5	smokers/non-smokers	1.3	55 (37–89)

Furthermore, in the EKA documentation a significant relationship between air exposure to ethylene oxide and the level of haemoglobin with the hydroxyethylvaline adduct is deduced on the basis of published data.

The MAK documentation for ethylene shows that, in the range of exposure concentrations which are important in practice, an external *ethylene* exposure leads to an internal formation of ethylene oxide which corresponds to an exogenous ethylene oxide exposure which is lower by a factor of 50. Therefore the following correlation between external occupational ethylene exposure and hydroxyethyl-adduct formation with the haemoglobin can be deduced, and is set as the EKA.

External exposure ppm ethylene	Hydroxyethylvaline in erythrocytes nmol/l Hb	μg/l blood
25	1.8	45
50	3.6	90
100	7.2	180
150	10.9	260

7 Interpretation of the Data

It should be borne in mind that smokers are subject to an exogenous ethylene/ethylene oxide exposure even without occupational exposure (see Table 1 and MAK documentation for ethylene and ethylene oxide).

Further, reference is made to the EKA documentation on ethylene oxide.

8 References

1. Andersen, M. E., Gargas, M. L., Jones, R. A., Jenkins Jr., L. J.: Determination of the kinetic constants for metabolism of inhaled toxicants in vivo using gas uptake measurements. Toxicol. Appl. Pharmacol. *54* (1980), 100–116
2. Bailey, E., Brooks, A. G. F., Dollery, C. T., Farmer, P. B., Passingham, B. J., Sleightholm, M. A., Yates, D. W.: Hydroxyethylvaline adduct formation in haemoglobin as a biological monitor of cigarette smoke intake. Arch. Toxicol. *62* (1988), 247–253
3. Bolt, H. M., Filser, J. G.: Kinetics and disposition in toxicology. Example: Carcinogenic risk estimate for ethylene. Arch. Toxicol. *60* (1987), 73–76
4. Brugnone, F., Perbellini, L., Faccini, G. B., Pasine, F., Bartolucci, G. B., DeRosa, E.: Ethylene oxide exposure. Biological monitoring by analysis of alveolar air and blood. Int. Arch. Occup. Environ. Health *58* (1986), 105–112
5. Calleman, C. J., Ehrenberg, L., Jansson, B., Osterman-Golkar, S., Segerbäck, D., Svensson, K., Wachtmeister, C. A.: Monitoring and risk assessment by means of alkyl groups in hemoglobin in persons occupationally exposed to ethylene oxide. J. Environ. Pathol. Toxicol. *2* (1978), 427–442
6. Conkle, J, P., Camp, B. J., Welch, B. E.: Trace composition of human respiratory gas. Arch. Environ. Health *30* (1975), 290–295
7. Denk, B.: Abschätzung des kanzerogenen Risikos von Ethylen und Ethylenoxid für den Menschen durch Speziesextrapolation von der Ratte unter Berücksichtigung der Pharmakokinetik. GSF-Bericht 20/90, GSF-Forschungszentrum für Umwelt und Gesundheit GmbH, München 1990
8. ECETOC: Ethylene oxide toxicology and its relevance to man: an up-dating of Technical Report No. 5. Technical Report Nr. 11, ECETOC, Brüssel 1984
9. Ehrenberg, L., Hiesche, K. D., Osterman-Golkar, S., Wennberg, I.: Evaluation of genetic risks of alkylating agents: Tissue doses in the mouse from air contaminated with ethylene oxide. Mutat. Res. *24* (1974), 83–103
10. Ehrenberg, L., Osterman-Golkar, S., Segerbäck, D., Svensson, K., Calleman, C. J.: Evaluation of genetic risks of alkylating agents. III. Alkylation of hemoglobin after metabolic conversion of ethene to ethylene oxide in vivo. Mutat. Res. *45* (1977),175–184

11. Farmer, P. B., Bailey, E., Gorf, S. M., Törnqvist, M., Osterman-Golkar, S., Kautiainen, A., Lewis-Enright, D. P.: Monitoring human exposure to ethylene oxide by the determination of haemoglobin adducts using gas chromatography-mass spectrometry. Carcinogenesis *7* (1986), 637–640
12. Filser, J. G., Bolt, H. M.: Inhalation pharmacokinetics based on gas uptake studies VI. Comparative evaluation of ethylene oxide and butadiene monoxide as exhaled reactive metabolites of ethylene and 1,3-butadiene in rats. Arch. Toxicol. *55* (1984), 219–223
13. Filser, J. G., Denk, B., Törnqvist, M., Kessler, W., Ehrenberg, L.: Pharmacokinetics of ethylene in man. Body burden with ethylene oxide and hydroxyethylation of hemoglobin due to endogenous and environmental ethylene. Arch. Toxicol. *66* (1992), 157–163
14. Föst, U.: Chemisch-analytische Möglichkeiten für ein "Biologisches Monitoring" bei Exposition gegenüber Vinylchlorid und Ethylenoxid. Dissertation, Universität Dortmund 1991
15. Harrison, V. C.: Ethylene, an ovulatory hormone? Lancet *21* (1981), 438 ff.
16. Hogstedt, B., Bergmark, E., Törnqvist, M., Osterman-Golkar, S.: Chromosomal aberrations and micronuclei in lymphocytes in relation to alkylation of hemoglobin in workers exposed in ethylene oxide and propylene oxide. Hereditas *113* (1990), 133–138
17. Mayer, J., Warburton, D., Jeffrey, A. M., Pero, R., Walles, S., Andrews, L., Toor, M., Latriano, L., Wazneh, L., Tang, D., Tsai, W.-Y., Kuroda, M., Perera, F.: Biologic markers in ethylene oxide-exposed workers and controls. Mutat. Res. *248* (1991), 163–176
18. Osterman-Golkar, S., Ehrenberg, L., Segerbäck, D., Hällström, I.: Evaluation of genetic risks of alkylating agents. II. Haemoglobin as a dose monitor. Mutat. Res. *34* (1976), 1–10
19. Osterman-Golkar, S., Farmer, P. B., Segerbäck, D., Bailey, E., Calleman, C. J., Svensson, K., Ehrenberg, L.: Dosimetry of ethylene oxide in the rat by quantitation of alkylated histidine in hemoglobin. Teratogen. Carcinogen. Mutagen. *3* (1983), 395–405
20. Passingham, B. J., Farmer, P. B., Bailey, E., Brooks, A. G. F., Yates, D. W.: 2-Hydroxyethylation of haemoglobin in man. In: Barsch, H., Hemminki, K., O'Neill, I. K. (eds.): Methods for detecting DNA-damaging agents in humans. IARC Scientific Publications No. 89. IARC, Lyon 1988, pp. 279–285
21. Ram Chandra, G., Spencer, M.: A micro-apparatus for absorption of ethylene and its use in determination of ethylene in exhaled gases from human subjects. Biochem. Biophys. Acta *69* (1963), 423–425
22. Sagai, M., Ichinose, T.: Age-related changes in lipid peroxidation as measured by ethane, ethylene, butane and pentane in respired gases of rats. Life Sci. *27* (1980), 731–738
23. Sarto, F., Törnqvist, M., Tomanin, R., Bartolucci, G. B., Osterman-Golkar, S. M., Ehrenberg, L.: Studies of biological and chemical monitoring of low-level exposure to ethylene oxide. Scand. J. Work Environ. Health *17* (1991), 60–64
24. Schmiedel, G., Filser, J. G., Bolt, H. M.: Rat liver microsomal transformation of ethene to oxirane in vitro. Toxicol. Lett. *19* (1983), 293–297
25. Segerbäck, D.: Alkylation of DNA and hemoglobin in the mouse following exposure to ethene and ethene oxide. Chem.-Biol. Interactions *45* (1983), 139–151
26. Shen, J., Kessler, W., Denk, B., Filser, J. G.: Metabolism and endogenous production of ethylene in rat and man. Arch. Toxicol., Suppl. *13* (1989), 237–239
27. Snellings, W. M., Weil, C. S., Maronpot, R. R.: A two-year inhalation study of the carcinogenic potential of ethylene oxide in Fischer 344 rats. Toxicol. Appl. Pharmacol. *75* (1984), 105–117
28. Snellings, W. M., Weil, C. S., Maronpot, R. R.: Final report on ethylene oxide two-year inhalation study on rats. Project report 44-20, Bushy Run Research Center (formerly Carnegie-Mellon Institute of Research) January 28, 1981. Submitted by Union Carbide Corporation to the US Environmental Protection Agency under Section 8 (e) of the Toxic Substances Control Act, on behalf of cosponsors of the study, 1981; quoted in ECETOC 1982 and 1984
29. Tates, A. D., Grummt, T., Törnqvist, M., Farmer, P. B., Van Dam, F. J., Van Mossel, H., Schoemaker, H. M., Osterman-Golkar, S., Uebel, Ch., Tang, Y. S., Zwinderman, A. H., Natarjan, A. T., Ehrenberg, L.: Biological and chemical monitoring of occupational exposure to ethylene oxide. Mutat. Res. *250* (1991), 483–497
30. Törnqvist, M.: Monitoring and cancer risk assessment of carcinogens, particularly alkenes in urban air. Dissertation, University of Stockholm (1989)
31. Törnqvist, M., Almberg, J. G., Bergmark, E. N., Nilsson, S., Osterman-Golkar, S.: Ethylene oxide doses in ethene-exposed fruit store workers. Scand. J. Work Environ. Health *15* (1989), 436–438

32. Törnqvist, M., Osterman-Golkar, S., Kautiainen, A., Jensen, S., Farmer, P. B., Ehrenberg, L.: Tissue doses of ethylene oxide in cigarette smokers determined from adduct levels in hemoglobin. Carcinogenesis *7* (1986), 1529–1521
33. Van Sittert, N., De Jong, G., Clare, M. G., Davies, R., Dean, B. J., Wren, L. J., Wright, A. S.: Cytogenetic, immunological, and haematological effects in workers in an ethylene oxide manufacturing plant. Br. J. Ind. Med. *42* (1985), 19–26
34. WHO: Ethylene oxide. Environ, Health Crit. *55* (1985)
35. Angerer, J., Schaller, K. H. in: Henschler, D., Greim, H. (eds.): Analytische Methoden zur Prüfung gesundheitsschädlicher Arbeitstoffe. Bd. 2: Analysen in biologischem Material. 7. Lfg. VCH Verlagsgesellschaft, Weinheim 1994
36. Lewalter, J.: unpublished investigations, Bayer AG, 1992

Author: H. M. Bolt
Approved by the Working Group: 19th November 1993

Ethylene oxide

EKA The correlation between external and internal exposure yields the following data:

Air ethylene oxide (ml/m³)	Air ethylene oxide (mg/m³)	Whole blood hydroxy-ethylvaline (µg/l)
0.5	0.92	45
1	1.83	90
2	3.66	180
3	5.49	260
4	7.32	350
5	9.15	440

Sampling: no restrictions

Date of evaluation 1993

MAK [last established: 1984] Carcinogenic substance: category IIIA2 of the List of MAK and BAT Values

1 Haemoglobin Adducts

Due to its electrophilic properties ethylene oxide reacts directly with cellular macromolecules (10, 4, 34, 39, 38). In addition to DNA adduct formation, in particular hydroxyethyl adducts of cysteine, histidine and the N-terminal valine in haemoglobin are described. The last have been suggested for biological monitoring of persons exposed to ethylene oxide (30, 11). These alkylation products are enriched by use of a modified Edman degradation of the haemoglobin (37).

2 Evaluation of Exposure Equivalents

Exposure equivalents for the formation of hydroxyethylvaline in human haemoglobin can be calculated with the formula given by Denk (5)

$$\text{HOEtVal [nmol/g Hb]} = \frac{P_{EO} \cdot t_{1/2}\,ery \cdot RC_{EO} \cdot 10^9}{\frac{\ln 2}{t_{1/2_{EO}}}}$$

where

P_{EO} is the uptake of ethylene oxide into the organism per kg body weight and day [mol/(kg · d)]
$t_{1/2}ery$ is the half-life of the erythrocytes [d] = 63
RC_{EO} is the reaction constant of the reaction of ethylene oxide with the N-terminal valine of haemoglobin [kg/(g · h)] = $5 \cdot 10^{-5}$
$t_{1/2_{EO}}$ is the half-life of ethylene oxide in the organism [h] = 42/60

A detailed calculation can be found in Filser et al. (13).

Denk (5) calculates a value of 42 minutes for the half-life of ethylene oxide in humans from the results of measurements which Brugnone et al. (3) carried out in employees exposed to ethylene oxide. Hattis (19) estimated a half-life of 41 minutes using the same data and a physiological pharmacokinetic model. Filser and Bolt (12) obtained a similar value of 39 minutes for humans using an allometric extrapolation of the half-life of 6 minutes determined in the rat. In rats, Filser and Bolt (12) found that 40 % of the substance inhaled per respiratory minute volume was cleared from the lung per minute. Observations with other readily water-soluble gases (5) indicate that this value can be applied to humans which means that 40 % of the ethylene oxide inhaled per unit of time is absorbed by the organism. Under these conditions and on the assumption of a half-life of 42 minutes (5), a steady state constant of 2.7 can be calculated for the accumulation of ethylene oxide in the organism. Correspondingly, Brugnone et al. (3) measured a 2.5-fold accumulation of ethylene oxide in the blood compared to the concentration in the ambient air.

From this data, exposure equivalents for hydroxyethylvaline in the haemoglobin of 1.8–18.1 nmol/g or 45–440 µg/l blood could be calculated using the formula given by Törnqvist et al. (37) for an air concentration of 0.5–5 ml/m^3 and an alveolar air concentration of 0.12–1.2 ml/m^3, assuming that 1 litre blood contains approx. 150 g haemoglobin (Tab. 1).

Tab. 1: EKA values for ethylene oxide

Air Ethylene oxide		Alveolar air Ethylene oxide		Blood HOEtVAL	
(ml/m^3)	(mg/m^3)	(ml/m^3)	(mg/m^3)	(nmol/g haemoglobin)	(µg/l)
0.5	0.92	0.12	0.22	1.8	45
1	1.83	0.24	0.44	3.6	90
2	3.66	0.48	0.88	7.2	180
3	5.49	0.72	1.32	10.9	260
4	7.32	0.96	1.76	14.5	350
5	9.15	1.2	2.2	18.1	440

For P_{EO} the formula $P_{EO} = C \cdot Q_{alv} \cdot 40\ \%$ was used where C_{alv} represents the concentration of ethylene oxide in the alveolar air. Q_{alv} was assumed to be 288 l kg^{-1} · d^{-1} and an exposure of 40 hours/week was taken as a basis.

Prerequisite for the theoretical derivation of these exposure equivalents is the notion of a linear function between the internal ethylene oxide dose and the external ethylene oxide exposure concentration. This has been demonstrated for the rat up to concentrations of 100 ppm (12, 19), and for humans up to at least 8 ppm by the investigations of Brugnone et al. (3). In inhalation studies with rats a linear relationship between exposure concentrations (0–110 ppm) and the level of haemoglobin adducts was also found (35, 31). The hydroxyethylation of haemoglobin bears a direct relationship to the hydroxyethylation of DNA (9, 31, 11, 29, 33). After short-term exposure in the concentration range between 1 and 33 ppm the hydroxyethylation of haemoglobin and of the liver DNA of the rat was directly proportional to the air concentration (33). With constant exposure concentration a direct proportionality between exposure duration and hydroxyethylation of the guanine was found (14).

More recent findings by Föst et al. (15), which suggest an enzyme dimorphism of the glutathione-S-transferase in human erythrocytes, call into question the validity of extrapolation to humans of functions deduced in animal experiments for the haemoglobin binding of ethylene oxide. Whether the level of internal exposure in individuals with slow adduct formation may be underestimated has yet to be clarified.

3 Relationships to Carcinogenic Risk

The linear relationship given also corresponds to linear relationships in the given concentration range between external exposure and carcinogenic risk. Estimation of this risk was carried out for the mutagenic effects of ethylene oxide by Latarjet et al. (25), by Ehrenberg et al. (9) and by Kolman et al. (24) by means of a so-called rad-equivalent approach. More recent estimations use a rad-equivalence factor deduced from empirical data from Denk (5) and from Denk and Filser (6). The authors also calculate the cancer risk allometrically by transfer of observations in rats to humans. For exposure of humans to 1 ppm ethylene oxide 40 hours per week for 45 years the cancer risk was estimated allometrically as 1.04 per 100 exposed persons. Using the rad-equivalence factor the risk for this exposure was estimated as being between 1.7 and 8.5 per 100 exposed persons. Corresponding to the notion of a linear dependence between risk and air concentration there is for an airborne concentration of 3 ppm a 3-fold higher cancer risk. The cancer risk from ethylene oxide in humans was also determined using a rad-equivalence factor by the research group associated with Ehrenberg (38). The risk estimation yielded a 20-fold higher value than that calculated by Denk (5) and Denk and Filser (6). The differing values are due to different assumptions for the rad-equivalence factors and for the half-life of ethylene oxide in the body. Not only the well-founded derivation of the factors used by Denk (5) and Denk and Filser (6) but also epidemiological observations by Hogstedt et al. (23) and Gardner et al. (16), which suggest an even lower cancer risk, speak for the somewhat greater plausibility of the risk assessment by Denk (5) and Denk and Filser (6). Four smaller epidemiological studies carried out predominantly with persons exposed to mixtures (21, 36, 22, 23) indicate that ethylene oxide induces malignant diseases of the lymphatic and haemotopoietic systems (1) and stomach carcinoma. These observations stand in contrast with the negative epidemiological findings of Morgan et al. (27), Gardner et al. (16), Hagmar et al. (20) and Wong et al. (40). The carcinogenic effect of ethylene oxide has been proved without doubt in animal experiments (7, 8, 35, 26, 17, 18, 28).

4 Methodology

The pentafluorophenylhydantoin derivative of hydroxyethylvaline produced in the modified Edman degradation according to Törnqvist et al. (37) is transformed into the trimethylsilyl ether derivative (32) and then determined selectively with the GC/MS technique on the mass m/z 440 (molecule-ion). The deuterated analogous compound serves as an internal standard.

5 Interpretation of the Data

N-Hydroxyethylvaline can also regularly be found as a haemoglobin adduct in persons not occupationally exposed. As the proved endogenous ethylene production in the organism does not fully explain the measured value (5), it must be assumed that ethylene is taken up from the environment in regionally greatly differing amounts. This has been analytically proved for industrialised regions (review in 5). Further uptake of ethylene has been demonstrated in cigarette smokers (37, 11, 32). Törnqvist et al. give 57 ± 25 pmol/g as the median value for non-smokers and 289 ± 138 pmol/g as the median value for smokers. Passingham et al. (32) describe a correlation between the formation of the hydroxyethylvaline adduct in haemoglobin and the number of cigarettes smoked daily. The highest value determined for 30 cigarettes/day was below the lowest given exposure level of 0.5 ml/m^3 ethylene oxide which led to an adduct level of 1.8 nmol/g Hb. It should nevertheless be taken into account in the interpretation of the data.

The correlation given does not apply for short-term exposures of a few hours or single days but to long-term activities with regularly recurring exposures. Greatly fluctuating exposures on different days can also be determined as the effect is cumulative. A definite sampling time is not necessary.

6 Conclusions

The applicability of the given exposure equivalents for biological monitoring has not been deduced directly from observations in humans but theoretically from experimental data in rats and humans. They correspond well, however, with data obtained experimentally in humans (20). Thus Bryant and Osterman-Golkar (2) give for an ethylene exposure level of 1 ppm a HOEtVal concentration of 2.4 nmol/g Hb. The exposure equivalents seem, however, as provisional equivalents to be sufficiently well-founded. They correlate roughly with the carcinogenic risk caused by exposure. They can only be used strictly with non-smokers as above all the uptake of ethylene in cigarette smoke leads to a higher internal exposure of the smoker to ethylene oxide. In small amounts ethylene is also formed endogenously and taken up from the environment. The determination of hydroxyethylvaline makes it possible to check exposure conditions at longer intervals of weeks or months. Whether a dimorphism of the erythrocytic glutathione transferase can result in underestimation of the internal exposure to ethylene oxide in individuals with slow adduct formation is at present awaiting clarification.

7 References

1. Bisanti, L., Maggini, M., Raschetti, R., Spila Alegiani, S., Menniti Ippolito, F., Caffari, B., Segnan, N., Ponti, A.: Cancer mortality in ethylene oxide workers. B. J. Ind. Med. *50* (1993), 317–324
2. Bryant, M. S., Osterman-Golkar, S.: Hemoglobin Adducts as Dosimeters of Exposure to DNA-Reactive Chemicals. CIIT *11* (1991), 1–12
3. Brugnone, F., Perbellini, L., Pasini, F., Bertolucci, G. B., DeRosa, E.: Ethylene oxide exposure: Biological monitoring by analysis of alveolar air and blood. Int. Arch. Occup. Environ. Health. *28* (1986), 105–112
4. Calleman, C. J., Ehrenberg, L., Jansson, B., Osterman-Golkar, S., Segerbäck, D., Svensson, K., Wachtmeister, C. A.: Monitoring and risk assessment by means of alkyl groups in hemoglobin in persons occupationally exposed to ethylene oxide. J. Environ. Pathol. Toxicol. *2* (1978), 427–442
5. Denk, B.: Abschätzung des kanzerogenen Risikos von Ethylen und Ethylenoxid für den Menschen durch Speziesextrapolation von der Ratte unter Berücksichtigung der Pharmakokinetik. Dissertation, GSF-Bericht 20/90, München 1990
6. Denk, B., Filser, J. G.: Abschätzung des durch Ethylen und Ethylenoxid bedingten kanzerogenen Risikos für den Menschen – Vergleich mit dem Risiko durch endogenes Ethylen. In: Schuckmann, R., Schopper-Jochum, S. (eds.): Berufskrankheiten, krebserzeugende Arbeitsstoffe, Biological Monitoring. 80. Jahrestagung der Deutschen Gesellschaft für Arbeitsmedizin e.V., 28.–31. Mai 1990, Gentner Verlag, Stuttgart 1991, pp. 397–401
7. Dunkelberg, H.: Carcinogenic activity of ethylene oxide and its reaction products 2-chloroethanol, 2-bromoethanol, ethylene glycol and diethylene glycol. I. Carcinogenicity of ethylene oxide in comparison with 1,2-propylene oxide after subcutaneous, administration in mice. (Ger.) Zbl. Bakt. Hyg. I. Abt. Orig. B. *174* (1981), 383–404
8. Dunkelberg, H.: Carcinogenicity of ethylene oxide and 1,2-propylene oxide upon intragastric administration to rats. Br. J. Cancer *46* (1982), 924–933
9. Ehrenberg, L., Hiesche, K. D., Osterman-Golkar, S., Wennberg, I.: Evaluation of genetic risks of alkylating agents: tissue doses in the mouse from air contaminated with ethylene oxide. Mutat. Res. *24* (1974), 83–103
10. Ehrenberg, L., Osterman-Golkar, S., Segerbäck, D., Svensson, K., Calleman, C. J.: Evaluation of genetic risks of alkylating agents. III. Alkylation of hemoglobin after metabolic conversion of ethene to ethene oxide in vivo. Mutat. Res. *45* (1977), 175–184
11. Farmer, P. B., Neumann, H.-G., Henschler, D.: Estimation of exposure of man to substances reacting with macromolecules. Arch. Toxicol. *60* (1987), 251–260
12. Filser, J. G., Bolt, H. M.: Inhalation pharmacokinetics based on gas uptake studies VI. Comparative evaluation of ethylene oxide and butadiene monoxide as exhaled reactive metabolites of ethylene and 1,3-butadiene in rats. Arch. Toxicol. *55* (1984), 219–233
13. Filser, J. G., Denk, B., Törnqvist, M., Kessler, W., Ehrenberg, L.: Pharmacokinetics of ethylene in man; body burden with ethylene oxide and hydroxyethylation of hemoglobin due to endogenous and environmental ethylene. Arch Tox. *66* (1992), 157-163

14. Föst, U., Marczynski, B., Kasemann, R., Peter, H.: Determination of 7-(2-hydroxyethyl)guanine with gas chromatography/mass spectrometry as a parameter for genotoxicity of ethylene oxide. Arch. Toxicol. Suppl. *13* (1989), 250–253
15. Föst, U., Hallier, E., Ottenwälder, H., Bolt, H. M., Peter, H.: Distribution of Ethylene Oxide in Human Blood and its Implications for Biomonitoring. Human & Experimental Toxicology *10* (1991), 25–31
16. Gardner, M. J., Coggon, D., Panett, B., Harris, E. C.: Workers exposed to ethylene oxide: a follow up study. Br. J. Ind. Med. *46* (1989), 860–865
17. Garman, R. H., Snellings, W. M., Maronpot, R. R.: Brain tumors in F-344 rats associated with chronic inhalation exposure to ethylene oxide. Neurotoxicology *6* (1985), 117–138
18. Garman, R. H., Snellings, W. M., Maronpot, R. R.: Frequency, size and location of brain tumours in F-344 rats chronically exposed to ethylene oxide. Food Chem. Toxicol. *24* (1986), 145–153
19. Hattis, D.: A pharmacokinetic/mechanism-based analysis of the carcinogenic risk of ethylene oxide. Center for Technology, Policy and Industrial Development at the Massachusetts Institute of Technology, Cambridge 1987
20. Hagmar L., Welinder, H., Linden, K., Attewell, R., Osterman-Golkar, S., Törnqvist, M.: An epidemiological study of cancer risk among workers exposed to ethylene oxide using hemoglobin adducts to validate environmental exposure assessments. Int. Arch. Occup. Environ. Health *63* (1991), 271–277
21. Hogstedt, C., Rohlen, O., Berndtsson, B. S., Axelson, O., Ehrenberg, L.: A cohort study of mortality and cancer incidence in ethylene oxide production workers. Br. J. Ind. Med. *36* (1979), 276–280
22. Hogstedt, C., Aringer, L., Gustavsson, A.: Ethylene oxide and cancer – Review of the literature and follow-up of two studies (Swed.) Arbete Hälsa *49* (1984), 1–32
23. Hogstedt, C., Aringer, L., Gustavsson, A.: Epidemiological support for ethylene oxide as a cancer causing agent. J. Am. Med. Assoc. *255* (1986), 1575–1578
24. Kolman, A., Segerbäck, D., Osterman-Golkar, S.: Estimation of the cancer risk of genotoxic chemicals by the rad-equivalence approach. In: Bartsch, H., Hemminki, K., O'Neill, I.K. (eds.) Methods for detecting in DNA damaging agents in humans: application in cancer epidemiology and prevention. IARC Scientific Publications, No. 89. IARC, Lyon 1988
25. Latarjet, R., Averbeck, D., Levy, S., Poirier, V.: Quantitative comparison of genotoxic effects of atomic energy and fossil-fuelled energy. Radequivalences for ethylene, ethylene oxide and formaldehyde – consequences for decisions at Government level. Proc. Int. Symp. on Health Impacts on Different Sources of Energy, International Atomic Energy Agency. No. IAEM-SM-254/49, Vienna 1981
26. Lynch, D. W., Lewis, T. R., Moorman. W. J., Burg, J. R., Groth, D. H., Khan, A., Ackerman, L. J., Cockrell, B. Y.: Carcinogenic and toxicological effects of inhaled ethylene oxide and propylene oxide. Toxicol. Appl. Pharmacol. *76* (1984), 69–84
27. Morgan, R.W., Claxton, K. W., Divine, B. J., Kaplan, S. D., Harris, V. B.: J. Occup. Med. *23* (1981), 767
28. National Toxicology Program: Toxicology and Carcinogenesis Studies of Ethylene Oxide (CAS No. 75-21-8) in B6C3Fl Mice (Inhalation Studies) (NTP TR 326; NIH Publ. No. 86-2582), Research Triangle Park, NC (1986)
29. Neumann, H. G.: Hemoglobin binding in control of exposure to and risk assessment of aromatic amines. In: Bartsch, H., Hemminki, K., O'Neill, I. K. (eds.): Methods for detecting DNA damaging agents in humans: applications in cancer epidemiology and prevention. IARC Scientific Publications, Nr. 89. IARC, Lyon 1988, pp. 157–165
30. Osterman-Golkar, S., Ehrenberg, L., Segerbäck, D., Hällstöm, I.: Evaluation of genetic risks of alkylating agents. II. Hemoglobin as a dose monitor. Mutat. Res. *34* (1976), 1–10
31. Osterman-Golkar, S.: Tissue doses in man: Implications in risk assessment. In: Hayes, A. W., Schnell, R. C., Miya, T. S. (eds.): Developments in the Science and Practice of Toxicology Elsevier, Amsterdam 1983, pp. 289–298
32. Passingham, B. J., Farmer, P. B., Bailey, E., Brooks, A. G. F., Yates, D. W.: 2-Hydroxyethylation of Hemoglobin in man. In: Bartsch, H., Hemminki, K., O'Neill, I. K. (eds): Methods for detecting DNA-damaging agents in humans: application in cancer epidemiology and prevention. IARC Scientific Publications, No. 89. IARC, Lyon 1988, pp. 279–285
33. Potter, D., Blair, D., Davies, R., Watson, W. P., Wright, A. S.: The relationship between alkylation of hemoglobin and DNA in Fisher 344 rats exposed to (^{14}C)ethylene oxide. Arch. Toxicol. Suppl. *13* (1989), 254–257
34. Segerbäck, D.: Alkylation of DNA and hemoglobin in the mouse after exposure to ethene and ethene oxide. Chem.-Biol. Interact. *45* (1983), 139

35. Snellings, W. M., Weil, C. S., Maronpot, R. R.: A two-year inhalation study of the carcinogenic potential of ethylene oxide in Fisher 344 rats. Toxicol. Appl. Pharmacol. *75* (1984), 105–117
36. Thiess, A. M., Frentzel-Beyme, R., Link, R., Stocker, W. G.: Mortality study on employees exposed to alkylene oxides (ethylene oxide/propylene oxide) and their derivatives. In: Prevention of Occupational Cancer – International Symposium. Occupational Safety and Health Series No. 46, International Labour Office, Geneva 1981, pp. 249–259
37. Törnqvist, M., Osterman-Golkar, S., Kautiainen, A., Jensen, S., Farmer, P. B., Ehrenberg, L.: Tissue doses of ethylene oxide in cigarette smokers determined from adduct levels in hemoglobin. Carcinogenesis *7* (1986), 39–42
38. Törnqvist, M., Gustafsson, B., Kautiainen, A., Harms-Ringdahl, M., Granath, F., Ehrenberg, L.: Unsaturated lipids and intestinal bacteria as sources of endogenous production of ethene and ethylene oxide. Carcinogenesis *10* (1989), 39–42
39. van Sittert, N. J., DeJong, G., Clare, M. G., Davies, R., Dean, B. J., Waren, L. J., Wright, A. S.: Cytogenetic, immunological, and hematological effects in workers in an ethylene oxide manufacturing plant. Br. J. Ind. Med. *42* (1985), 19–26
40. Wong, O., Trent, L. S.: An epidemiological study of workers potentially exposed to ethylene oxide. B. J. Ind. Med. *50* (1993), 308–316

Authors: K. Norpoth, H. M. Bolt
Approved by the Working Group: 5th May 1994

Hydrazine

EKA for exposure to hydrazine and its salts

The correlation between external and internal exposure yields the following data:

Air Hydrazine (ml/m³)	(mg/m³)	Urine[1] (µg hydrazine/g creatinine)	Plasma[1] Hydrazine (µg/l)
0.01	0.013	35	27
0.02	0.026	70	55
0.05	0.065	200	160
0.08	0.104	300	270
0.10	0.130	380	340

1) Sampling: end of exposure or end of shift

Date of evaluation 1991

Synonym Diamide

Formula NH_2NH_2 (water-free)
$NH_2NH_2\ (H_2O)_x$ (hydrates)
NH_2NH_3A; $NH_3NH_3A_2$ (salts, A = anion, acid residue)

Molecular weight 32.05

Melting point 1.8 °C

Boiling point 113.5 °C

Vapour pressure at 20 °C 1.3 hPa

MAK [last established: 1985] Carcinogenic substance: category IIIA2 of the List of MAK and BAT Values

Details of the metabolism and toxicology of hydrazine have been published by the DFG Commission for the Investigation of Health Hazards of Chemical Compounds in the Work Area (11).

1 Metabolism and Kinetics

Hydrazine (or hydrazine hydrate) is quickly absorbed by inhalation and through the skin (9, 25). It is excreted in urine mainly as the monoacetylated and diacetylated forms (13, 21, 35). An accumulation of hydrazine has so far not been observed in humans (14, 28). Hydrazine, particularly in toxic doses, causes an indirect methylation of the O^6 and N7 position of the deoxyguanosine of the liver DNA (experimental data from rat liver) (2, 14) (cf. Fig. 1).

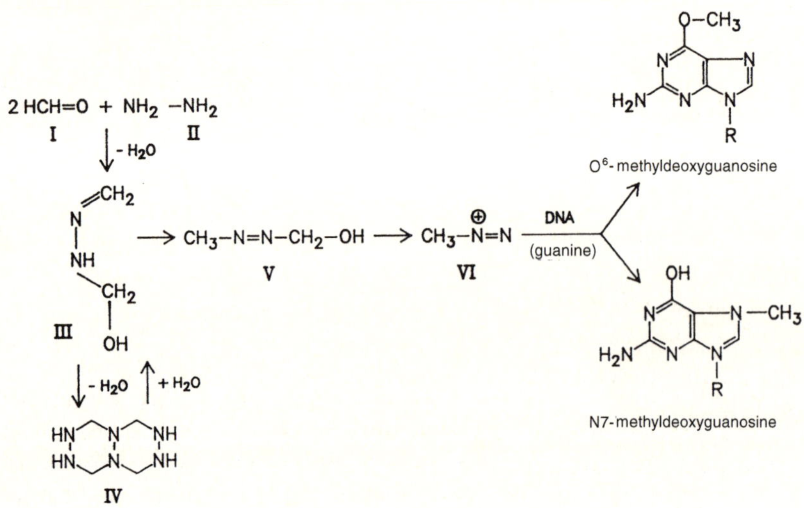

Fig. 1: Postulated mechanism of the DNA methylation by free hydrazine (26, 27).

Exogenous hydrazine (II) is thought to react with endogenous substances of the C1 metabolic pool, e.g., formaldehyde (I) bound to tetrahydrofolic acid, to form a formaldehyde-hydroxymethyl-hydrazone (III) (31). This intermediate (III) could then, in analogy to the alkylhydrazines, be activated enzymatically via methylazomethanol (V) to the electrophilic methyldiazonium ion (VI) which then methylates DNA bases (15). The resulting base adducts are N7-methyldeoxyguanosine and O^6-methyldeoxyguanosine (2, 23) as well as O^6-methylthymidine (17) and N3-methyldeoxyadenosine (8, 16).

Investigations into the alkylation of proteins show that as well as DNA other macromolecules are indirectly methylated by hydrazine. The nature of these adducts is so far unknown (21). Hydrazine-induced methylation of the N-terminal valines in the globin of haemoglobin could, however, not be demonstrated either by animal experiments nor in persons exposed to hydrazine (16).

It can be assumed that the individual acetylation capacity modifies the genotoxicity of hydrazine. Thus, in persons exposed to hydrazine, methylation of deoxyguanosine of the leucocyte DNA is then observed when the incorporated hydrazine is acetylated "slowly" and therefore apparently cannot be eliminated quickly enough (16).

2 Critical Toxicity

Acute intoxication with hydrazine in animal experiments resulted in kidney and liver damage (9, 33). Oral application of high hydrazine doses to rats and mice led to pulmonary adenomas, adenocarcinomas, hepatomas and hepatocarcinomas. After an intraperitoneal hydrazine dose lymphomas and leukaemia were reported (12).

With acute hydrazine intoxication central nervous effects with vomiting, muscular tremor, convulsions and paraesthesia but also liver and kidney damage were described (9). (For further details on the symptoms of acute intoxication see the MAK documentation for hydrazine (11)).

According to currently available data, a carcinogenic effect of hydrazine in humans after exposure to hydrazine as it occurs at the workplace cannot be demonstrated (for further details see the MAK documentation for hydrazine (11) – in English translation in the series "Occupational Toxicants", obtainable from VCH Verlagsgesellschaft mbH, Weinheim).

3 Exposure and Effects

External exposure is to be understood as one that can lead to absorption by inhalation or through the skin. The internal exposure is the biologically relevant concentration of the hydrazine or the products of its metabolism in the target tissue. As parameters for internal exposure the hydrazine concentrations in the plasma and urine can be used.

Methylated DNA bases can be regarded as parameters of internal hydrazine effects. The determination of the incidence of DNA strand breaks in leucocytes represents a further parameter of effect of internal hydrazine exposure (20, 31). As this parameter is very unspecific it is not recommended for biological monitoring.

4 Selection of the Indicators

Internal exposure to hydrazine and its effects can be determined through the hydrazine concentrations in urine and plasma or via hydrazine-induced methylation rates in the blood and liver cells.

4.1 Hydrazine in Urine and in Plasma

The hydrazine concentrations in plasma and excretion of hydrazine in urine observed after occupational contact with hydrazine have only recently been routinely determined. For this reason correlation investigations between inhaled hydrazine levels and the corresponding hydrazine concentration in plasma or urinary excretion of free and conjugated hydrazine are lacking in the literature. The data listed in this EKA documentation are based on examination of workers with continual occupational exposure to hydrazine.

Plasma is preferred for analysis in acute accident cases with hydrazine. Chronic, intermittent and retrospective hydrazine exposures are usually determined via urine excretion.

4.2 Hydrazine-induced DNA Methylation

In DNA the positions N7 and O^6 in guanine as well as N3 in adenine are the preferred targets for the methylation induced by hydrazine (2, 14, 16). The biological half-life of O^6-methyldeoxyguanosine is determined above all by the individual repair capacity (7, 24, 32). The N3-methyladenine formation which is independent of endogenous methylation can serve as a measure of the methylation potential of absorbed hydrazine (8, 16, 22, 29).

Biological monitoring is in principle possible via the determination of the ratio of N7-methylguanine to O^6-methylguanine in the leucocyte DNA as well as the N3-methyladenine excretion in urine (1, 34). To what extent the value of the N7-methylguanine to O^6-methylguanine ratio in leucocytes or the N3-methyladenine excretion in urine can be correlated with adverse effects needs further investigation (4, 6, 18, 19).

5 Methodology

The analysis of free and monoacetylated hydrazine takes place after derivatisation and extraction by capillary gas chromatography with an ECD detector or via GC/MS (gas chromatography/mass spectrometry) (17).

Methylation of the DNA of leucocytes can be analysed with a GC/MS (3, 29) or immunologically (5). N3-Methyladenine in urine is determined immunologically (8, 22, 30) or chromatographically (1). Further methods of investigation are described in the literature (10, 17).

The determination of the incidence of DNA strand breaks is carried out after alkaline elution of the DNA (20, 31).

6 Evaluation of the Biological Exposure Equivalents for Carcinogenic Substances (EKA)

The concentration of hydrazine in target tissues after inhalation of hydrazine and/or percutaneous hydrazine exposure cannot be determined directly. The pharmacokinetic situation is such that the hydrazine concentrations in urine and in plasma, the N3-methyladenine concentration in urine and the ratio of N7-methylguanine to O^6-methylguanine in the leucocyte DNA can serve as indicators of internal exposures and effects.

6.1 Exposure Parameters – Hydrazine in Urine and in Plasma

Occupational contact with hydrazine concentrations below the currently valid TRK value (0.1 ml hydrazine/m^3) leads to a dose-dependent increase in the hydrazine levels in the plasma and in the urine. Correlations between external and internal exposure can therefore be established for the evaluation of EKA values. In the literature such data are still not available. The Commission has at its disposal, however, extensive data from a company in the chemical industry (for the years 1978–1991). The most important results are summarised in table 1.

The table contains, in addition to the mean values, also information on variation. Particularly the internal exposure was very variable in some cases. The individual differences can be due to differing metabolism (acetylation and methylation) and differing routes of absorption (inhalation and percutaneous). Therefore both parameters should be simultaneously determined where possible.

Tab. 1: Relationship between the external concentration of hydrazine in the ambient air and the plasma and urine concentrations of hydrazine in workers with continual exposure to hydrazine (J. Lewalter, unpublished investigations)

Persons n	Air ml N_2H_4/m^3	Urine[1)] µg N_2H_4/g creatinine	Plasma[1)] µg N_2H_4/l
37[2)]	0.01	35 (5–70)	27 (15–45)
30[3)]	0.02	70 (25–120)	55 (40–65)
3	0.05	195 (150–210) = 200[4)]	160 (135–180)
2	0.08	300 (300; 305)	270 (255; 285)
–	0.10	380[5)]	340[5)]

[1)] Sampling: end of exposure or end of shift
[2)] 14 smokers
[3)] 9 smokers
[4)] Rounded-up value
[5)] Extrapolated data

6.2 Effect Parameters - Hydrazine-induced Methylation of DNA

After exposure to hydrazine concentrations at the currently valid TRK value (0.1 ml hydrazine/m^3), no additional methylation of the guanine in the leucocyte DNA was found, only the normally observed methylation levels. The N3-methyladenine excretion in urine is not affected in this range of exposure. This agrees with the lack of cytotoxic effects (9) and statistically inconspicuous incidence of DNA strand breaks (20, 31) at these hydrazine doses. For this reason no parameters of effect can be given as EKA values for exposure levels up to the TRK value valid at present.

7 Interpretation of the Data

For the reasons mentioned above, only the determination of hydrazine in plasma and in urine is suitable for monitoring persons with chronic occupational hydrazine exposure in the concentration range below the TRK value. After acute intoxication in addition to this determination in plasma and in urine, determination of the N3-methyladenine excretion in urine and the level of N7-methylation of guanine in the leucocytes is to be recommended. This monitors not only current exposures but also any exposures within the previous week.

The hydrazine determinations in urine and in plasma, as well as the N3-methyladenine determination in urine, must be carried out after the end of the shift or after an accident. The samples must be analysed immediately or at least stabilised by immediate reaction with the detection reagent. The breakdown of the hydrazine can also be prevented by immediate freezing. The hydrazine content in urine has been expressed in terms of the creatinine level as no other correlation data are available.

8 References

1. Autrup, H.: Human exposure to genotoxic carcinogens: methods and their limitations. J. Cancer Res. Clin. Oncol. *117* (1991), 6–12
2. Becker, R. A., Barrows, L. R., Shank, R.: Methylation of liver DNA guanine in hydrazine hepatoxicity: dose-response and kinetic characteristics of 7-methylguanine and O^6-methylguanine formation and persistence in rats. Carcinogenesis *2* (1981), 1181–1188
3. Bolt, H. M., Peter, H., Föst, U.: Analysis of macromolecular ethylene oxide adducts. Int. Arch. Occup. Environ. Health *60* (1988), 141–144
4. Bosan, W. S., Lambert, C. E., Shank, R. C.: The role of formaldehyde in hydrazine-induced methylation of liver DNA guanine. Carcinogenesis *1* (1986), 413–418
5. Degan, P., Montesano, R., Wild, Ch. P.: Antibodies against 7-methyldesoxyguanosine: Its detection in rat peripheral blood lymphocyte DNA and potential applications to molecular epidemiology. Cancer research *48* (1988), 5065–5070
6. ECETOC: DNA and protein adducts: Evaluation of their use in exposure monitoring and risk assessment. ECETOC Monograph Nr. 13, Brüssel 1989
7. Doerjer, G., Bedell, M. A., Oesch, F.: DNA adducts and their biological relevance. In: Obe, G. (ed.): Mutation in man (1984), 20–34
8. Eberle, G., Glüsenkamp, K.-H., Drosdziok, W., Rajewsky, M. F.: Monoclonal antibodies for the specific detection of 3-alkyladenines in nucleic acids and body fluids. Carcinogenesis *11* (1990), 1753–1759
9. Environmental Health Criteria 68: Hydrazine. WHO, Geneva 1987
10. Gustavson, S., Proper, J. A., Bowie, W. E., Sommer, St. S.: Parameters affecting the yield of DNA from human blood. Anal. Biochem. *165* (1987), 294–299
11. Henschler, D. (ed.): Gesundheitsschädliche Arbeitsstoffe. Toxikologisch-arbeitsmedizinische Begründung von MAK-Werten. VCH Verlagsgesellschaft, Weinheim (1972, 1986)
12. International Agency for Research on Cancer (IARC): The evaluation of the carcinogenic risk of chemicals to humans. IARC Monograph No. 7. IARC, Lyon 1974, pp. 37–72
13. Kaneo, Y., Iguchi, S., Kubo, H., Iwagird, N., Matsuyama, K.: Tissue distribution of hydrazine and its metabolites in rats. J. Pharm. Dyn. *7* (1984), 556–562
14. Kimball, R. F.: The mutagenicity of hydrazine and some of its derivatives. Mutat. Res. *39* (1977), 111–126
15. Lambert, C. E., Shank, R. C.: Role of formaldehyde hydrazone and catalase in hydrazine-induced methylation of DNA guanine. Carcinogenesis *9* (1988), 65–70

16. Lewalter, J.: Personal experience during the biological monitoring of persons exposed to hydrazine (1978–1990)
17. Henschler, D. (ed.): Analytische Methoden zur Prüfung gesundheitsschädlicher Arbeitsstoffe, Bd.2: Analysen in biologischem Material, Hydrazin-Bestimmung per Kapillargaschromatographie und ECD. In preparation. VCH Verlagsgesellschaft, Weinheim
18. Lutz, W. K.: Quantitative evaluation of DNA binding data for risk estimation and for classification of direct and indirect carcinogens. J. Cancer Res. Clin. Oncol. *112* (1986), 85–91
19. Lutz, W. K.: Structural characteristics of compounds that can be activated to chemically reactive metabolites: use for a prediction of a carcinogenic potential. Arch. Toxicol. Suppl. *7* (1984), 194–207
20. Parodi, S., De Flora, S., Gavanna, M., Pino, A., Robbiani, L., Bennicelli, C., Brambilla, G.: DNA-damaging activity in vivo and bacterial mutagenicity of sixteen hydrazine derivates as related quantitatively to their carcinogenicity. Cancer Res. *41* (1981), 1469–1482
21. Perry, T. L., Kish, S. J., Hansen, S., Wright, J. M., Wall, R. A., Dunn, W. L., Bellward, G. D.: Elevation of brain GABA content by chronic low-dosage administration of hydrazine, a metabolite of isoniazid. J. Neurochem. *37* (1981), 32–39
22. Prevost, V., Shuker, D. E. G., Bartsch, H., Pastorelli, R., Stillwell, W. G., Trudel, L. J., Tannenbaum, S. R.: The determination of urinary 3-methyladenine by immunoaffinity chromatography - monoclonal antibody-based ELISA: use in human biomonitoring studies. Carcinogenesis *11* (1990), 1747–1751
23. Roggers, K. J., Pegg, A. E.: Formation of O^6-methylguanine by alkylation of rat liver, colon and kidney DNA following administration of 1,2-dimethylhydrazine. Cancer Res. *37* (1977), 4082–4087
24. Rüdiger, H. W., Lehnert, G.: Toxicogenetic: Grundlagen, Methoden und Bedeutung für die Arbeitsmedizin. ASP-Sonderheft *11* (1988), 4–16
25. Schmidt, E. W.: Hydrazine and its derivatives, preparation, properties, applications. Wiley & Sons, New York 1984
26. Shank, R. C., Bosan, W. S., Lambert, C. E.: Studies on the mechanism of action of hydrazine-induced methylation of DNA guanine. National Technical Information Service, AFAMRL-TR-85-057, Springfield, USA, 1984
27. Shank, R. C., Bosan, W. S., Lambert, C. E.: Role of formaldehyde in hydrazine-induced methylation of liver DNA guanine. National Technical Information Service, AAMRL-TR-85-052, Springfield, USA, 1985
28. Shank, R. C.: Evidence for indirect genetic damage as methylation of DNA guanine in response to cytotoxicity. Dev. Toxicol. Environ. Sci. *11* (1983), 145–152
29. Shuker, D. E. G., Bailey, E., Gorf, S. M., Lamb, J., Farmer, P. B.: Determination of N-7-(^{2}H3)methylguanine in fat urine by gas chromatography-mass spectrometry following administration of trideuteromethylating agents or precursors. Anal. Biochem. *140* (1984), 270–275
30. Shuker, D. E. G., Bailey, E., Parry, A., Lamb, J., Farmer, P. B.: The determination of urinary 3-methyladenine in humans as a potential monitor of exposure to methylating agents. Carcinogenesis *8* (1987), 959–962
31. Sina, J. E., Bean, C. L., Dysart, G. R., Taylor, V. I., Bradley, M. O.: Evaluation of the alkaline elution/rat hepatocyte assay as a predictor of carcinogenic/mutagenic potential. Mutat. Res. *113* (1983), 357–391
32. Topal, M. D.: DNA repair, oncogenes and carcinogenesis. Carcinogenesis *9* (1988), 691–696
33. Vernot, E. H., Macewen, J. D., Bruner, R. H., Haun, C. C., Kinkead, E. R., Prentice, D. E., Hall, A., Schmidt, R. E., Eason, R. L., Hubbard, G. B., Young, J. T.: Long-term inhalation toxicity of hydrazine. Fundam. Appl. Toxicol. *5* (1985), 1050–1064
34. Wild, Ch. P.: Antibodies to DNA alkylation adducts as analytical tools in chemical carcinogenesis. Mutat. Res. *233* (1990), 219–233
35. Wright, J. M., Timbrell, J. A.: Factors affecting the metabolism of ^{14}C-acetylhydrazine in rats. Drug Metab. Dispos. *6* (1978), 561–556

Author: J. Lewalter
Approved by the Working Group: 3rd May 1991

Nickel and sparingly soluble nickel compounds

(Nickel as nickel metal, nickel sulphide, sulphidic ores, nickel oxide, nickel carbonate)

EKA for exposure to nickel and sparingly soluble nickel compounds

The correlation between external and internal exposure yields the following data:

Air μg nickel/m^3	Urine μg nickel/l
100	15
300	30
500	45

Date of evaluation 1990

Chemical symbol Ni

Atomic weight 58.69

Melting point 1453 °C

Boiling point 2732 °C

MAK Carcinogenic substance: category IIIA1 of the List of MAK and BAT Values

A series of reviews and monographs have recently appeared on the occurrence, uptake, distribution and toxicity of nickel and its compounds (23, 25, 28, 30, 34, 39, 43, 46).

1 Metabolism and Kinetics

The gastro-intestinal and broncho-pulmonary systems are the main routes of absorption of nickel and its compounds. Uptake via the skin and also via parenteral routes (endoprostheses etc.) is of markedly less importance.

In the non-occupational area by far the largest amount of nickel is taken up with food. Although individual foods, among other things due to their methods of preparation, contain extraordinarily different concentrations of this metal it can be assumed that an average adult takes in approx. 300 µg of nickel daily with food (46). Only a small amount of this, however, given at values between 1 % and 4 %, is absorbed and excreted via the kidneys. The rest is in the faeces. With a low-protein diet or under fasting conditions considerably more nickel can be absorbed.

Assuming that in urban areas the nickel concentrations in the air are on average 20 ng/m^3, then the daily uptake of this metal by inhalation is considerably less than the oral uptake.

At *workplaces exposed to nickel,* the relationship between the two routes of uptake is reversed. Here inhalation is of much greater importance than uptake via the gastro-intestinal tract, not only due to the amount but also due to the toxic effect. The site of deposition of the inhaled nickel dusts depends upon the size of the dust particles. With decreasing aerodynamic diameter (<10 µm) an increasing proportion of the dust particles reach the alveoli. The larger dust particles are deposited in the nose, throat, laryngeal area and tracheobronchial tree. Part of the deposited dust is exhaled again or is taken up via the gastrointestinal tract after clearance from the lung. The rest is absorbed in the lung. The rate of absorption depends on the physical-chemical properties of the nickel compound involved, in particular on the water solubility, the crystal form and size, the surface charge, etc. Readily soluble nickel compounds, such as nickel(II) chloride, are quickly absorbed and over 90 % is excreted via the kidneys (8, 13). The half-time of nickel elimination from the plasma and urine is 1–1.5 days, as was ascertained in the investigation of workers who were exposed to nickel in an electroplating works (44). Elimination half-times of this order of magnitude have also been described for stainless-steel welders (1). These authors have also demonstrated that in exposed welders, in addition to this relatively fast elimination, the nickel is excreted with a second considerably longer half-time. Even after a 4-week holiday these authors could determine significantly increased nickel concentrations in the urine of welders. With decreasing solubility of the nickel compounds these are absorbed from the lung more slowly and more incompletely. After occupational exposure to sparingly soluble nickel compounds, as for example in the smelting of nickel ores, half-times of nickel elimination of up to several years have been found. This means that sparingly soluble nickel compounds accumulate to a considerable extent in the bronchopulmonary system. Even in persons not occupationally exposed the nickel content of the lungs shows a marked increase with age (35, 36, 45).

Nickel is distributed very evenly in the human organs. Only in the lungs, thyroid gland and adrenal gland are there slightly higher values than for example in the kidneys, heart, liver, brain etc. (36). The nickel concentrations in lung tissue are quoted at values between 20 and 40 µg/kg wet weight (35, 45).

The nickel concentrations in the body fluids, blood and urine, described as normal values have shown a decreasing tendency over the last 10 years (2). This is mainly to be attributed to the increasing sophistication of the analytical methods and not, however, to a decreased uptake of nickel. It can be assumed that the nickel concentration in blood and plasma is normally below 1 µg/l. The upper normal limit for the nickel concentration of urine is about 3 µg/l (2, 17, 27). This value corresponds well with the concentrations to be expected from the uptake with food.

After occupational exposure to nickel, accumulation of the metal has not been observed in organs other than the bronchopulmonary system. However, the concentrations in blood and urine increase considerably.

The transport of nickel in the human body takes place mainly in plasma, seldom, however, in the cells of the blood such as, e.g., in the erythrocytes. In plasma the nickel is bound to two high-molecular fractions (albumin and α_2-macroglobulin) and a fraction which can be ultrafiltered.

2 Critical Toxicity

In the last ten years a series of reviews and monographs on the toxicology of nickel have been published (7, 15, 16, 28, 29, 31, 37, 40).

After occupational exposure to nickel and its compounds, from the point of view of occupational medicine the carcinogenic and allergic effects are the most important.

The toxic effects of nickel and its compounds depend greatly on the way in which they entered the body. While orally absorbed nickel shows no systemic toxicity, after inhalation of this metal and its compounds the carcinogenic effect on the respiratory tract is relevant.

2.1 Carcinogenicity

At the workplace the inhalation of this metal and its compounds is of the greatest importance. The proved or potential carcinogenic effect of the various nickel species is the most important toxicological aspect. The target organs of the carcinogenic effect are above all the nasal cavities and paranasal sinuses as well as the bronchial tract. In investigations of large collectives who were exposed in Norway, England and Canada during the hot-dip smelting of nickel ores and the subsequent refining, it was proved that certain work processes were associated with an excessive cancer risk of the localities named above (11, 12, 24, 32, 38). Sunderman (43) has discussed various hypotheses on the carcinogenic effects of nickel.

Occupational-medical experience complemented by data from animal experiments and experiments with cell cultures have caused the Commission for the Investigation of Health Hazards of Chemical Compounds in the Work Area to classify "nickel (in the form of inspirable dusts/aerosols of nickel metals, nickel sulphide and sulphidic ores, nickel oxide and nickel carbonate, as they can occur during production and processing)" in category A1 with substances which have been shown to induce malignant tumours in humans (10). For readily soluble nickel compounds there is to date no evidence of human carcinogenicity.

2.2 Nickel-Induced Allergenic Effects

So-called nickel scabies, contact dermatitis occurring after sensitisation, is the longest known occupational toxicological illness caused by nickel. The development of this allergy is – given endogenous reactivity – dependent on the exposure concentration of the substance. The syndrome in itself, once allergy has developed, is not dependent on concentration. The eczema occurs at the site of exposure so that its cause can usually be determined. In over 80 % of cases sensitisation has its origins in private life as a consequence of skin contact with the metal or its compounds, e.g. in the form of fashion accessories, buttons etc. It is assumed that there is a relationship between the clinical activity of nickel dermatitis and the level of the nickel concentration in blood and urine (21, 22).

An oversensitivity to nickel can also cause asthmatic illnesses, conjunctivitis and inflammatory reactions round implants and endoprostheses (43).

3 Selection of the Indicators

Above all in the selection of the indicators the determination of the nickel concentration in the urine and plasma should be discussed.

For many reasons, urine is the preferred medium for determination of indicators. The sampling of urine for routine monitoring is easier to carry out than the taking of blood. Furthermore, the sampling of urine causes less inconvenience for the test person. But above all, urine has in its favour that the nickel concentration in this medium is approx. 8–10 times higher than that in blood. Thus this parameter is diagnostically considerably more sensitive than the determination in blood or serum. In addition to the better diagnostic sensitivity of the urine level as a parameter, at present nickel can also be determined in urine with greater analytical sensitivity. With the extraction-enrichment processes usual today and atomic absorption spectrometry, nickel concentrations in the physiological and environmental range can be determined. With these methods, concentrations as low as 0.1 µg nickel per litre urine can be routinely determined (3, 27).

Furthermore, the database for the external nickel exposure and the resulting nickel excretion in the urine is very extensive. Compared to this there are only few data for the relationship between the nickel concentration in the air, on the one hand, and that in the serum, on the other.

Urine has disadvantages, however, compared to serum, where a different uptake of fluids greatly changes the nickel concentration in urine. Also the danger of exogenous contamination endangers the accuracy of the nickel findings in urine. In these cases, in particular when the confirmation of individual values is required, therefore, in addition to the nickel concentration in urine the metal concentration in serum should be determined. Whole blood seems less suitable as it is now known that nickel does not occur in the erythrocytes.

4 Methodology

For the determination of nickel the Working Group "Analysen in biologischem Material" ("Analyses of Hazardous Substances in Biological Materials") of the DFG Commission for the Investigation of Health Hazards of Chemical Compounds in the Work Area has published a tested analytical procedure (3, 4).

In the procedure for the determination of nickel in urine the metal is separated from the urine by means of a chelation/extraction step and at the same time concentrated. The quantitative determination takes place by atomic absorption spectrometry after electrothermal atomisation in the graphite furnace. The evaluation is carried out with aqueous standard solutions which are prepared and analysed in the same way. This procedure is simple and can be carried out in a large series. Therefore it is particularly suited for routine monitoring. The high sensitivity of the method combined with the low detection limit enables concentrations not only in the occupational medical range but also the general environmental concentration range to be determined accurately. The detection limit is approx. 0.1 µg nickel/l urine. As the total sample preparation can be carried out in a disposable container and requires only few procedural steps and small quantities of chemicals, the error due to losses and/or endogenous contamination can be reduced to a minimum. The low susceptibility to interference is reflected in high precision of the procedure. The specificity of the procedure is a result of the extraction step during sample preparation and the specificity of the atomic absorption spectrometric method. The procedure suggested by the IUPAC for the determination of nickel in the urine or serum (6) also uses a chelation/extraction step. Recent experience indicates that the mineralisation of the biological material stipulated there can be dispensed with.

For the determination of nickel in whole blood or serum an atomic absorption spectrometric procedure has proved suitable in which the biological material after eightfold dilution is analysed directly and without further sample preparation. The biological sample is subjected to a three-stage mineralisation programme in a graphite furnace before the atomisation takes place at 2650 °C. Evaluation takes place by means of the so-called standard addition procedure in which increasing, known amounts of nickel are added to the sample which is analysed each time.

Using this direct procedure the influence of the matrix on the analysis result is minimised, on the one hand, by dilution of the biological material, on the other, by the greatly differentiated mineralisation procedure. The detection of nickel can therefore take place very specifically. A minimum analytical background interference is eliminated without difficulty by the use of deuterium compensation or the Zeeman effect. Due to the dilution of the sample, the detection limit is between 1 and 2 µg/l so that physiological or environmentally-produced nickel concentrations in the blood or serum cannot be determined. The procedure is, however, suitable without restrictions for the occupational medical concentration range. The accuracy and precision of the method are good. Other procedures described in the literature for the determination of nickel in serum or whole blood which use de-proteinisation (41) or mineralisation (6) or in which oxygen (9) or a matrix modifier is used as a mineralisation aid, produce poorer results (18).

5 Relationship between External and Internal Nickel Exposure and Evaluation of Exposure Equivalents

In the following synopsis of the literature available on the question of a relationship between external and internal nickel exposure only those investigations are included in which humans are in contact with sparingly soluble nickel compounds. As already mentioned, these compounds are characterised by their carcinogenic effect. Furthermore, the kinetics of the uptake and excretion of all these compounds are similar. Biological monitoring can be carried out on this basis.

Table 1 shows a summary of the studies in which the external and internal nickel exposure at different workplaces have been reported. In all cases the exposure was to sparingly soluble forms of nickel. At the workplaces where nickel alloy steel is welded (1, 2, 14, 33, 47) it is to be assumed that the welding fumes contain nickel mostly in metallic and oxidised form. There is exposure to metallic nickel in the mechanical processing of nickel alloys by grinding and polishing (5). Also in flame spraying (5, 34) and the production of metal powders (26) the exposure to metallic nickel is foremost. On the other hand in the drying of metal oxides (26) or the roasting of metal sulphides (19) there is exposure to nickel oxides and sulphides.

A comparison of the results of the studies listed in table 1 makes clear that there is nothing that speaks against the view that these different nickel species are subject to comparable uptake and excretion kinetics. This view is supported by investigations by various authors who have succeeded in producing good correlations between the nickel concentration in the air at the workplace and the nickel concentration in the urine of the exposed persons (14, 33, 35, 42) (Tab. 2). In these studies the values measured in the ambient air are results of personal air sampling. The results by Emmerling et al. (14) whose relationship is based on the investigation of 176 welders seem to be particularly useful in this respect. Sunderman et al. (42) have succeeded in creating a relationship between the nickel concentration in air and that in urine on the basis of groups with different exposures. This correlation is based on results reported by Morgan and Rouge (26).

From these observations the exposure equivalents represented in table 3 for nickel concentrations in urine can be deduced. They refer firstly to urine as the assay material. This parameter is of the greatest importance for monitoring exposed collectives for the reasons mentioned above (see section 3). The orienting values given for the nickel concentration in plasma were calculated from the known relationship between nickel concentrations in plasma and urine; according to this the ratio of the nickel concentration in urine to that in the serum is about 8:1 (2, 5).

Tab. 1: Relationship between the concentration of sparingly soluble nickel compounds in the air at the workplace and the nickel concentration in the blood and urine of the exposed persons

$\bar{x}$: mean
s: standard deviation
m: median
–: no data
M: male
F: female

MAG: metal active gas welding
MMA: manual metal arc welding
TIG: tungsten inert gas welding

No.	Occupation	Test persons	Exposure concentration Air µg/m³ $\bar{x}$ ± s; (range)	m	Plasma/serum µg/l $\bar{x}$ ± s; (range)	m	Urine µg/l $\bar{x}$ ± s; (range)	m	Authors
1	Polishers of aeroplane parts	6M 1F	26 ± 41; (0.05 – 129.0)	–			4.1 ± 3.2; (0.5 – 9.5)	–	Bernacki et al., 1978
	Grinders (nickel alloys)	7M 2F	1.6 ± 3.0; (0.02 – 9.5)	–			5.4 ± 2.4; (2.1 – 8.8)	–	
	Welders (MMA)	7M 3F	6.0 ± 14.3; (0.2 – 46.0)	–			6.3 ± 4.1; (1.6 – 14.0)	–	
	Aviation mechanics	4M 4F	52.0 ± 94.0; (0.01 – 252.0)	–			12.2 ± 13.6; (1.4 – 41.0)	–	
	Flame sprayers	4M 1F	2.4 ± 2.6; (0.04 – 6.5)	–			17.2 ± 9.8; (1.4 – 26.0)	–	
	Nickel refining								
2	Roasting, melting	24	860.0 ± 1200.0;	–	7.2 ± 2.8;	–	65.0 ± 58.0;	–	Høgetveit et al., 1978
	Other areas	13	420.0 ± 490.0;	–	6.4 ± 1.9;	–	45.0 ± 27;	–	
3	Nickel refining; (Exposure to insoluble nickel compounds)		240.0;	–			31.7;		Høgetveit et al., 1980

* 68 %-range

Tab. 1: Continued

No.	Occupation	Test persons	Exposure concentration Air µg/m³ $\bar{x}$ ± s; m (range)	Plasma/serum µg/l $\bar{x}$ ± s; m (range)	Urine µg/l $\bar{x}$ ± s; m (range)	Authors
4	Welders (high alloyed stainless steel; up to 34 % Ni) (MMA)	10M	150.0; – (30.0 – 1780.0)	1.2(B); – (0.58 – 2.9)(B)	11.5*; – (7.8 – 26.5)	Rahkonen et al., 1983
5	Nickel catalyst production	73	(<200.0 – 5870.0)	8.0; – (2.0 – 41.0)	64.0; – (9.0 – 300.0)	Zwennis and Franssen, 1983
6	Stainless steel welders (TIG)	10M	–; 2.3 (0.7 – 9.0)		–; 4-6 (1.4 – 9.5)	Zober et al., 1984
	Smelting of ores containing nickel					
7	Drying of nickel oxides	67	310; – (10.0 – 5000)		24.0 ± 24.0; –	Morgan and Rouge, 1984
	Drying of nickel oxides	37		8.9 ± 5.9; –	37.0 ± 30.0; –	
	Production of nickel powder	48	310; – (90.0 – 1530)			
	Production of nickel powder	25		7.2 ± 4.8; –		
	Contact with sparingly soluble nickel compounds	15	1540; – (220 – 4180)		39 ± 28; –	
		7		7.4 ± 5.1; –		

8	Welding of high alloyed steel (75 % Ni)	11M	440; – (70.0 – 1100)		18; – (9.7 – 38.0)	Åkesson and Skerfing, 1985
9	Welders (stainless steel) (MMA)	41M	(95 % < 500)		–; 23.8 (2.5 – 144.0)	Zober and Weltle, 1985
	Hollow glass industry-mould manufacturing + maintenance					
10	Mechanics	406	(18 – 3800)		7.4; – (2.9 – 24.3)	Raithel et al., 1987
	Mechanics	140		0.9; – (0.75 – 2.05)		
	Powder flame sprayers	114	(3 – 600)		25.3; – (8.5 – 81.5)	
	Powder flame sprayers	40		1.95 (0.75 – 3.25)		
	Combined jobs	394	(300 – 410)		17.5 (4.9 – 53.9)	
	Combined jobs	108		1.65 (0.75 – 4.10)		
11	Stainless steel welders (MAG, MMA)	103M	93 ± 81.4; 60 (<50 – 320)	4.9 ± 4.0; 3.9 (<1.8 – 19.6)	18.5 ± 28.5; 10.2 (0.1 – 209.4)	Angerer and Lehnert, 1989
	Stainless steel welders					
12	MMA	61	–; 24 (7.4 – 52.7)*	–; 1.0 (0.4 – 3.2)*	–; 10.1 (3.4 – 20.1)*	Emmerling et al., 1989
	MAG	46	–; 68.6 (22.1 – 238.7)*	–; 1.5 (0.7 – 2.8)	–; 14.0 (4.4 – 34.3)*	
	TIG	16	–; 8 (2.1 – 20.3)*	–; 1.1 (0.2 – 2.7)	–; 5.7 (1.8 – 15.4)*	

Tab. 2: Correlation between the concentration of sparingly soluble nickel compounds in the air of the workplace and the metal concentration in the body fluids c_U, c_A: nickel concentration in the urine and air

Linear regressions		Number of pairs of value	Correlation coefficient	Authors
c_U (µmol/l)	$= 0.08 + 0.8\ c_A$(mg/l);	10	0.95	Rahkonen et al., 1983
c_U (µg/g creatinine)	$= 13.5 + 0.05\ c_A$(µg/m³);	22	0.52	Raithel, 1987
c_U (µg/l)	$= 10.84 + 0.007\ c_A$(µg/m³);	174	0.52	Emmerling et al., 1989
c_U (µg/l)	$= 8.49 + 56.88\ c_A$(mg/m³);		0.86	Sunderman et al., 1986* according to data by Morgan and Rouge, 1984

* Correlation on the basis of groups exposed to different extents

Tab. 3: Exposure equivalents for nickel valid for exposures to sparingly soluble nickel compounds

Air µg nickel/m³	Urine µg nickel/l	Plasma µg nickel/l
	(EKA values)	(supplementary values for purposes of orientation)
100	15	<2
300	30	4
500	45	6

6 Interpretation of the Data

The exposure equivalents evaluated in section 5 for nickel are only valid for the occupational medical surveillance of exposed persons who are exposed to sparingly soluble nickel compounds, such as nickel sulphide, nickel oxides or metallic nickel in powder form. This applies above all to activities which are connected with the mining of ores containing nickel, their smelting and refining, with the mechanical processing (e.g. grinding) of alloys containing nickel, with the welding of alloys containing nickel and with the so-called powder flame spraying.

The exposure equivalents do *not* apply for contact with readily soluble nickel salts, for exposure to nickel carbonyl or for the area of electrolysis of solutions containing nickel salts.

The kinetics of the uptake and elimination of nickel described in section 1 indicated that the sampling of biological material should take place at the end of the work period after several consecutive work shifts.

As is generally the case after exposure to heavy metals, exogenous contamination of biological material can lead to considerable errors in the values measured. Care must therefore be taken that sampling takes place in a contamination-free atmosphere. The test person should as far as possible be wearing ordinary clothes. As already mentioned, high *individual values* in the analysis of urine samples should be verified by additional determination of the nickel content of plasma samples, better still by the renewed investigation of a urine sample.

The different dilution of urine samples is taken into account most easily if the specific weight of the urine is used as an exclusion criterion. Urines with specific weights <1,010 are not included in the evaluation.

As already mentioned, the normal or ecologically-defined upper normal limits of the nickel concentration in urine or plasma lie at around 3 µg/l urine or under the detection limit of approx. 1 µg/l plasma. Higher concentrations of nickel indicate an occupational exposure.

In the case of nickel concentrations in urine which are less than 10 µg/l it can be assumed with great probability that the exposure of the person in question is permanently less than the currently valid TRK value (status 1990). The prerequisite is, however, that the urine is of normal concentration. The specific weight should be markedly greater than 1.01.

7 References

1. Åkesson, B., Skerfing, S.: Int. Arch. Occup. Environ. Health *56* (1985), 111–117
2. Angerer, J., Lehnert, G.: Int. Arch. Occup. Environ. Health *62* (1990), 7–10
3. Angerer, J., Schaller, K. H.: Analyses of Hazardous Substances in Biological Materials. Vol. 1. Deutsche Forschungsgemeinschaft. VCH Verlagsgesellschaft, Weinheim 1985
4. Angerer, J., Schaller, K. H.: Analyses of Hazardous Substances in Biological Materials. Vol. 3. Deutsche Forschungsgemeinschaft. VCH Verlagsgesellschaft, Weinheim 1990
5. Bernacki, E. J., Parsons, G. E., Roy, B. R., Mikac-Devic, M., Kennedy, C. D., Sunderman, F. W.: Ann. Clin. Lab. Sci. *8* (1978), 184–189
6. Brown, S. S., Nomoto, S., Stoeppler, M., Sunderman, F. W. jr.: Clin. Biochem. *14* (1981), 295–299
7. Brown, S. S., Sunderman, F. W. (eds.): Progress in nickel toxicology. Blackwell, Oxford 1985
8. Carvalho, S. M. M., Ziemer, P. L.: Arch. Environ. Contamin. Toxicol. *11* (1982), 245–248
9. Delves, H. T. et al.: Proceedings of the 2nd international workshop trace element analytical chemistry, Neuherberg, April 1982, Vol. 2. de Gruyter, Berlin 1983, pp. 1123–1127
10. Deutsche Forschungsgemeinschaft: Maximale Arbeitsplatzkonzentrationen und Biologische Arbeitsstofftoleranzwerte 1990. Mitteilung XXVI der Senatskommission zur Prüfung gesundheitsschädlicher Arbeitsstoffe. VCH Verlagsgesellschaft, Weinheim 1990
11. Doll, R., Morgan, L. G., Speizer, F. E.: Br. J. Cancer *24* (1970), 623–632
12. Doll, R., Mattheus, J. D., Morgan, L. G.: Br. J. Ind. Med. *34* (1977), 102–105
13. English, J. C., Parker, R. D.: Am. Ind. Hyg. Assoc. J. *42* (1989), 486–492
14. Emmerfing, G., Zschiesche, W., Schaller, K. H., Weltle, D., Valentin, H.: Arbeitsmedizinische Untersuchung von Chrom-Nickel-Stahlschweißern: Epidemiologische Querschnittsstudie zur Belastung sowie zur bronchopulmonalen und renalen Beanspruchung. DVS-Verlag, Düsseldorf 1989
15. Environmental Protection Agency (EPA): Relative carcinogenic potencies among 54 chemicals evaluated by the carcinogen assessment group as suspect human carcinogens 11 124 11 126
16. Grandjean, Ph., Selikoff, I. J., Shen, S. K., Sunderman, F. W.: Am. J. Ind. Med. *1* (1980), 181–189
17. Heinrich, R., Angerer, J. in: Bolt, H. M., Piekarski, C., Rutenfranz, J. (eds.): 25. Jahrestagung in Dortmund vom 22. bis 25. Mai 1985. Gentner Verlag, Stuttgart 1985
18. Heinrich-Ramm, R., Angerer, J.: Trace elements in human health and disease, in: WHO, CEC, EPA (eds.) Environmental Health *20* (1987), 70–73
19. Høgetveit, A. C., Barton, R. T., Kostal, C. O.: Ann. Occup. Hyg. *21* (1978), 113-120
20. Høgetveit, A. C., Barton, R. T., Andersen, J.: J. Occup. Med. *22* (1980), 597–600

21. Jongh, De, G. J., Spruit, D. et al.: Contact Dermatitis *4* (1978), 142–156
22. Kaaber, K., Veien, N.: Br. J. Dermatol. *98* (1978), 197–201
23. Leonard, A. et al.: Mutat. Res. *87* (1981), 1–15
24. Magnus, K., Andersen, A., Høgetveit, A. C.: Int. J. Cancer *30* (1982), 681–685
25. Mastromatteo, E.: Am. Ind. Hyg. Assoc. J. *47* (1986), 589–601
26. Morgan, L. G., Rouge, P. J. C. in: Sunderman, F. W. (ed.): Nickel in the human environment. IARC Lyon 1984, pp. 507–520
27. Mundt-Becker, W., Angerer, J.: Int. Arch. Occup. Environ. Health *49* (1981), 187–192
28. Nieboer, E., Rossetto, F. E., Menon, C. R., In: Sigel, H., Sigel, A. (eds.): Metal ions in biological systems, Vol. 23. Marcel Dekker, New York 1988
29. National Institute for Occupational Safety and Health (NIOSH): NIOSH Publication No. 77–164. US Government Printing Office, Washington DC 1977, pp. 1–282
30. Norseth, T., Piscator, M.: Nickel. In: Friberg, F., Nordberg, G. F., Vouk, V. B. (eds.): Handbook on the toxicology of metals. Elsevier, Amsterdam 1979
31. Nriagu, J. O. (ed.): Environ. Sci. Technol. Serv. Wiley and Sons, New York 1980
32. Pedersen, E., Høgetveit, A. C., Andersen, A.: Int. J. Cancer *12* (1973), 32–41
33. Rahkonen, E., Junttila, M.-L., Kalliomäki, P. L., Olkiroura, M., Koponen, M., Kalliomäki, K.: Int. Arch. Occup. Environ. Health *52* (1983), 243–255
34. Raithel, H. J., Ebner, G., Schaller, K. H., Schellmann, B., Valentin, H.: Am. J. Ind. Med. *12* (1987), 55–70
35. Raithel, H. J.: Arbeitsmed. Sozialmed. Präventivmed. *22* (1987), 268–274 and 301–310
36. Rezuki, W. N., Knight, J. A., Sunderman jr., F. W.: Am. J. Ind. Med. *11* (1987), 419
37. Rigaut, J. P.: Commission des Communautes Europeennes. Luxembourg Do CCE/LUX/V/E/24/83 (1983)
38. Shannon, H. S., Julian J. A., Roberts, R. S.: JNCI. *73* (1984), 1251–1258
39. Stockinger, H. E.: Nickel. In: Clayton, G. D., Clayton, F. E. (eds.): Patty's Industrial Hygiene and Toxicology, 3rd ed. Vol. 2. Wiley & Sons, New York 1981
40. Sunderman, F. W. (ed.): IARC Scientific Publications, No. 53, Lyon 1984
41. Sunderman, F. W. et al.: Annals of clinical and laboratory science *14* (1984), 232–241
42. Sunderman, F. W., Aitio, A., Morgan, L. G., Norseth, T.: Toxicol. Ind. Hlth. *2* (1986), 17–78
43. Sunderman, F. W. In: Seiler, H. G., Sigel, H., Sigel, A. (eds.): Handbook on toxicity of inorganic compounds. Marcel Dekker, New York 1988
44. Tossavainen, A., Nurminen, M., Mutanen, P., Tola, S.: Br. J. Ind. Med. *37* (1980), 285–291
45. Turhan, V., Wollburg, C., Angerer, J., Szadkowski, D.: Arbeitsmed. Sozialmed. Präventivmed. *20* (1985), 277–281
46. Zimmerli, B., Candrian, U., Schlatter, Ch.: Mitt. Gebiete Lebensm. Hyg. *78* (1987), 344–396
47. Zober, A., Schaller, K. H., Weltle, D.: Staub Reinh. Luft *44* (1984), 465–468
48. Zober, A., Weltle, D.: J. Soc. Occup. Med. *35* (1985), 79–84
49. Zwennis, W. C. M., Franssen, A. G.: Abs. Second Intern. Conf. Clin. Chem. Clin. Toxicol. Metals, Montreal (1983)

Author: J. Angerer
Approved by the Working Group: 28th March 1990

Vinyl chloride

EKA

The correlation between external and internal exposure yields the following data:

Air Vinyl chloride		Urine Thiodiglycolic acid
(ml/m^3)	(mg/m^3)	(mg/24 h)
1	2.6	1.8
2	5.2	2.4
4	10	4.5
8	21	8.2
16	41	10.6

Sampling: 24 hours after the beginning of the shift

Date of evaluation 1986

Synonyms: Chloroethylene
VCM

Formula: C_2H_3Cl

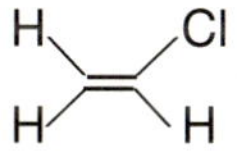

Molecular weight: 62.50

Melting point: –159.7 °C

Boiling point: –13.9 °C

MAK [last established: 1977]: Carcinogenic substance: category IIIA1 of the List of MAK and BAT Values

1 Metabolism and Kinetics

1.1 Metabolism

Vinyl chloride (VCM) is, after uptake into the mammalian organism, either exhaled unchanged or converted after oxidative biotransformation into metabolites which are excreted renally. In figure 1 the metabolism is shown according to the currently available data (12, 23, 28).

In animal experiments carboxymethylcysteine, thiodiglycolic acid (thiodiacetic acid) and hydroxyethylmercapturic acid can be detected as degradation products. The metabolic pathway which leads to thiodiglycolic acid can be regarded as certain, while for the formation of hydroxyethylmercapturic acid various pathways are possible. Which of these pathways are followed can at present not be decided. Investigations on employees exposed to VCM show that thiodiglycolic acid is quantitatively the most important metabolite (22). Carboxymethylcysteine is an intermediate which is degraded

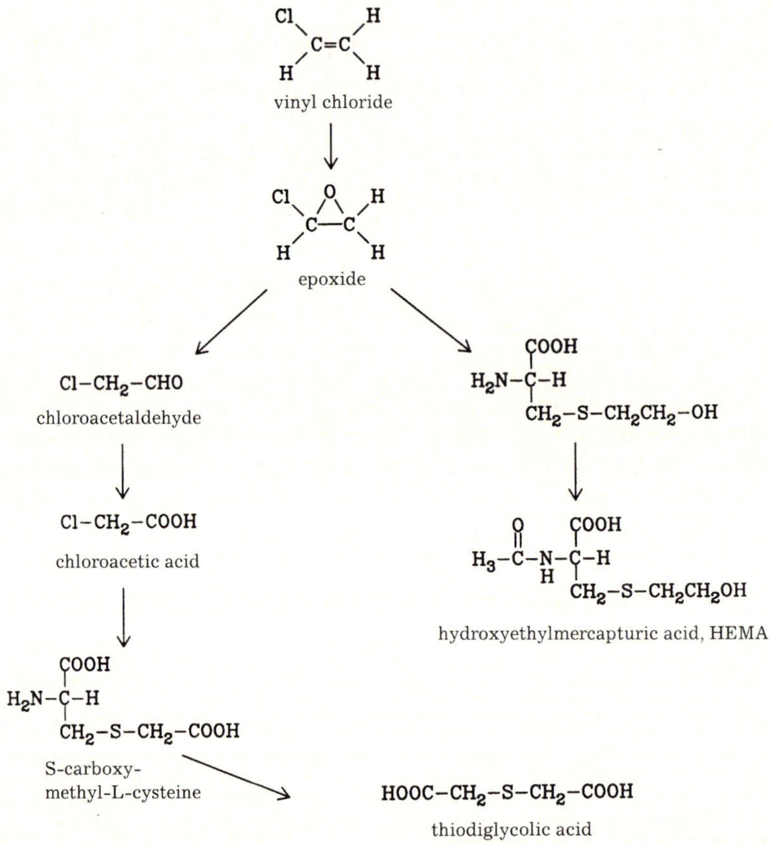

Fig. 1: Simplified diagram of the metabolism of vinyl chloride

to thiodiglycolic acid while the hydroxyethylmercapturic acid is only formed and excreted in small amounts, possibly because the enzyme activity of the microsomal glutathionyl transferases in the human liver is one tenth of that in rat liver (5).

1.2 Kinetics

Vinyl chloride is absorbed from the air relatively rapidly. With constant level exposure an equilibrium develops after approximately 30 minutes between the VCM concentration in the air and the VCM concentration in the body. Under these conditions approximately 25 % of the VCM amount inhaled is degraded by the organism through metabolism (6). The equilibrium constant for the steady state between the concentrations of VCM in the air and the body in persons of normal weight is 0.8 and increases in overweight persons as a consequence of the increased amount of adipose tissue to 1.2 (14). According to this the VCM concentration in the body can be calculated as follows:

body VCM concentration =

$$\left(\frac{0.8 \times \%\ \text{non-adipose tissue} + 0.8 \times \%\ \text{adipose tissue}}{100}\right) \cdot \text{air VCM concentration.}$$

The amount of VCM taken up and metabolised fluctuates and is dependent upon the changes in exposure profile with time. Thus, continuous low-level exposure leads to a higher internal exposure of the organism than comparable short-term peak exposures. This has been shown in model calculations (4, 9) and also in investigations of employees exposed to VCM (14).

2 Critical Toxicity

The symptoms of disease attributed to exposure to vinyl chloride in man are grouped together and called “vinyl chloride disease“ (18). Originally acro-osteolysis, Raynaud phenomena, scleroderma-like changes of the skin, restrictive lung function disorders and liver function disorders were included in this syndrome. Further symptoms are thrombocytopenia, endocrine changes (hyperthyroidism, sexual disorders), immunological changes (hyper-immunoglobulinaemia with increased IgE, cryoglobulinaemia) (11). Also neurological symptoms were reported in workers exposed to vinyl chloride (17). The intensity of scleroderma-like changes induced by VCM seems to be associated with an accumulation of HLA-antigens (2). The changes mentioned are understood as direct tissue toxicity of vinyl chloride or its metabolites and as secondary changes due to vascular and/or immunological processes (11, 17). The actual critical effects of vinyl chloride are its human carcinogenic effects. The target organs of the carcinogenic effects include the liver, brain, lungs, hematopoietic and lymphatic systems (16). An increased incidence of tumours for liver angiosarcoma could be verified in many epidemiological investigations (3, 10, 13, 26, 27). For the development of liver angiosarcomas a latency period of between 5 and 40 years is given. Moreover, an increased incidence of brain tumours in workers exposed to vinyl chloride (20, 26, 29) and an increased incidence of lung tumours (7, 13, 15, 20, 26, 29) has been reported. Recently and for the first time a slightly increased incidence of malignant melanomas and thyroid gland carcinomas (15) has been detected. These findings must be clarified. In rats the formation of liver angiosarcomas, hepatomas, carcinoma of the Zymbal gland, nephroblastomas, neuroblastomas, mammary gland adenocarcinomas, forestomach papillomas and acanthomas and extrahepatic angiosarcomas could be shown (19). Tumour development after exposure to concentrations below 5 ppm or 0.03 mg/kg vinyl chloride was not detected.

Results for DNA alkylation by reactive VCM metabolites in several tissues (1) are in accordance with the report of a multipotent carcinogenic effect of vinyl chloride. It has become known that the introduction of etheno groups into DNA bases causes miscoding (30). Causal relationships between DNA binding of VCM metabolites and carcinogenesis, however, have yet to be clarified.

3 Exposure and Effects

Data from a study on the relationship between the exposure to the toxic substance (determined by personal air sampling) and the thiodiglycolic acid excretion (14) are available. The analysis of the data shows that it is not possible as a matter of course to deduce relationships between the VCM concentrations in the air and the concentration of the thiodiglycolic acid in the urine as individual variables, for example, the body weight, have considerable influence on the excretion of the metabolites. If, however, with the help of a kinetic model, one calculates the VCM concentrations in the

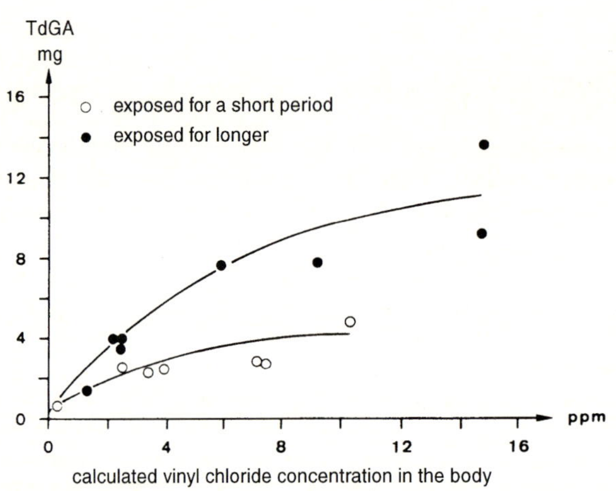

Fig. 2: Increase in the amount of thiodiglycolic acid (TdGA) in the urine of persons exposed to vinyl chloride (VCM), collected in the 24-hour period after the beginning of the shift, plotted against the effective concentration of VCM in the body calculated on the basis of the VCM concentrations in air determined by personal air monitoring (12-hour average value)

body which build up after uptake of the substance (14) and determines the relationship between the values obtained and the data for the thiodiglycolic acid excretion in the urine then plausible functions result for persons exposed for long periods and persons exposed for short periods at the same level of exposure (Fig. 2).

4 Selection of the Indicators

The only parameter available at present for the evaluation of the risk for persons exposed to VCM is thiodiglycolic acid in urine. For other urinary metabolites such as for example hydroxyethylmercapturic acid (22) and carboxymethylcysteine (21), dependence on the VCM uptake which would prove their validity as monitoring parameters, has not been demonstrated.

5 Methodology

5.1 Thiodiglycolic Acid in Urine

The determination of the thiodiglycolic acid in urine takes place by gas chromatography/mass spectrometry of the methyl ester (mass 146) after derivatisation with diazomethane and with the standard addition method (21). The method of determination of the thiodiglycolic acid as trimethylsilylester described by Draminski and Trojanowska (8) does not fulfil the validity criteria required for persons exposed to VCM due to its detection limit of 10 mg thiodiglycolic acid/l of urine.

5.2 Normal Values of Thiodiglycolic Acid

Normal values of thiodiglycolic acid can at present only be determined using the GC/MS-technique. For men these amount to 0.67 ± 0.32 mg/l and for women 0.51 ± 0.20 mg/l (24). It can be deduced from these investigations that a differentiation between exposed persons and non-exposed persons is possible when the 99 % tolerance level of 1.82 mg /l for men and 1.43 mg/l for women is exceeded.

Substances which also lead to the excretion of thiodiglycolic acid are listed in Tab. 1. These are workplace substances or drugs from which in the course of biotransformation chloroacetaldehyde can be produced or which have structures so similar to that of VCM that they yield the same end product (e.g. acrylonitrile) (25).

Tab. 1: Workplace substances and drugs which, like vinyl chloride, are transformed to the urinary metabolite, thiodiglycolic acid

Workplace substances	Drugs
Acrylonitrile	BCNU (Carmustine)
2,2´-Bis(chloroethyl)ether	CCNU (Lomustine)
Bromoethanol	Cyclophosphamide
Chloroethanol	Dibenamine
1,2-Dibromoethane	Dibenzyline
1,2-Dichloroethane	Distraneurin®
1,1,2-Trichloroethane	Iphosphamide
Vinyl bromide	Mannomustine

6 Evaluation of the Biological Exposure Equivalents for Carcinogenic Substances (EKA)

For a medium body weight of 75 kg relationships can be deduced from the data published between the VCM concentration in air and the thiodiglycolic acid excretion over 24 hours after the beginning of the shift. This correlation between the external exposure and the excretion of the metabolite is given in the following table.

Air Vinyl chloride (ml/m^3)	Urine Thiodiglycolic acid (mg/24 h)
1	1.8
2	2.4
4	4.5
8	8.2
16	10.6

7 Interpretation of the Data

The metabolism of vinyl chloride is subject to considerable individual fluctuations. Influencing factors include the degree of physical work and the body weight (see section 1.2). The thiodiglycolic acid excretion is influenced by drugs and other substances which are also transformed into this metabolite (see Tab. 1). To date little is known about the significance of individual differences in the activity of the microsomal oxidation systems.

The monitoring of persons exposed to VCM on the basis of the determination of thiodiglycolic acid in the urine requires the collection of all urine samples in the period of 24 hours from the beginning of the workshift. Single samples taken at the end of the shift cannot be recommended for investigation as the excretion of the metabolites in such samples is subject to excessive fluctuation relative to the VCM concentrations in air. Also the determination of concentrations in the urine cannot be recommended at present as only for the amount of metabolite excreted in 24 hours is the relationship to the VCM concentration in air known with sufficient certainty.

8 References

1. Bergman, K.: Arch. Toxicol. *49* (1982), 117–129
2. Black, C. M., Walker, A. E., Catoggio, T. J., Welsh, K. I., Bernstein, R. M., McGregor, A. R., Lloyd Jones, J. K.: Lancet *I* (1983), 53–55
3. Block, J. B.: JAMA *229* (1974), 53–54
4. Bolt, H. M., Filser, J. G., Buchter, A.: Arch. Toxicol. *48* (1981), 213–228
5. Bolt, H. M., Filser, J. G., Oesch, F., Guenthner, T. M., Friedberg, T., Bolt, M. In: Verhandlungen der Deutschen Gesellschaft für Arbeitsmedizin, 18. Jahrestagung. Gentner, Stuttgart 1981, pp. 507–509
6. Buchter, A., Bolt, H. M., Filser, J., Goergens, H. W., Laib, R. J., Bolt, W. In: Verhandlungen der Deutschen Gesellschaft für Arbeitsmedizin, 18. Jahrestagung. Gentner, Stuttgart 1978, pp. 111–124
7. Buffler, P. A., Wood, S., Eifler, C., Suarez, L., Kilian, D. J.: J. Occup. Med. *21* (1979), 195–202
8. Draminski, W., Trojanowska, B.: Arch. Toxicol. *48* (1981), 289–291
9. Filser, J. G., Bolt, H. M.: In: Verhandlungen der Deutschen Gesellschaft für Arbeitsmedizin. 22. Jahrestagung. Gentner, Stuttgart 1982, pp. 631–635
10. Fox, A. J., Collier, P. E.: Br. J. Ind. Med. *34* (1977), 1–10
11. Granger, R. G., Walker, A. E., Ward, A. M. In: Preger, L. (ed.): Drug, Irradiation, Occupation. Grune & Stratton 1979, pp. 191–124
12. Green, T., Hathway, D. E.: Chem. Biol. Interactions *II* (1975), 545–562
13. Greiser, E., Reinl, W., Weber, H.: Zentralbl. Arbeitsmed. *32* (1982), 44–62
14. Heger, M., Müller, G., Norpoth, K.: Int. Arch. Occup. Environ. Health *50* (1982), 187–196
15. Heldaas, S. S., Langard, S. L., Anderen, A.: Br. J. Ind. Med. *41* (1984), 25–30
16. International Agency for Research on Cancer (IARC): The evaluation of the carcinogenic risk of chemicals to humans. Chemicals, industrial processes and industries associated with cancer in humans, IARC Monographs Suppl. 4. IARC, Lyon 1982
17. Langauer-Lewowicka, H., Kurzbauer, H., Byczkowska, Z., Wocka-Marek, T.: Int. Arch. Occup. Environ. Health *52* (1983), 151–157
18. Lange, C. E., Jühe, S., Stein, G., Veltman, G.: Int. Arch. Occup. Health *32* (1974), 1–32
19. Maltoni, C., Lefemine, G., Ciliberti, A., Cotti, G., Carretti, D.: Environ. Health Persp. *41* (1981), 3–29
20. Monson, R. R., Peters, J. M., Johnson, M. N.: Lancet *II* (1974), 397–398
21. Müller, G., Brzezinka, H.: Thiodiessigsäure (Bis-(carboxymethyl)sulfid). In: Henschler, D. (ed.): Analytische Methoden zur Prüfung gesundheitsschädlicher Arbeitsstoffe. Bd. 2: Analysen in biologischem Material. VCH Verlagsgesellschaft, Weinheim 1982
22. Müller, G., Heger, M., Norpoth, K. In: Verhandlungen der Deutschen Gesellschaft für Arbeitsmedizin. 20. Jahrestagung. Gentner, Stuttgart 1980, pp. 533–536
23. Müller, G., Norpoth, K. In: Verhandlungen der Deutschen Gesellschaft für Arbeitsmedizin. 17. Jahrestagung. Gentner, Stuttgart 1977, pp. 237–241
24. Müller, G., Norpoth, K., Wickramasinghe, R. H.: Int. Arch. Occup. Environ. Health *44* (1979), 185–191
25. Müller, G., Verkoyen, C., Soton, N., Norpoth, K.: Arch. Toxicol. *60* (1987), 464–466
26. Tabershaw, I. R., Gaffey, W. R.: J. Occup. Med. *16* (1974), 509–518
27. Thériault, G., Allard, P.: J. Occup. Med. *23* (1981), 671–676
28. Watanabe, P. G., McGrowan, G. R.: Toxicol. Appl. Pharmacol. *36* (1976), 339–352
29. Waxweiler, R. J., Stringer, W., Wagoner, J. K., Jones, J., Falk, H., Carter, C.: Ann. N. Y. Acad. Sci. *271* (1976), 40–48
30. Zajdela, F., Croisy, A., Barbin, A., Malaville, C., Tomatis, L., Bartsch, H.: Cancer Res. *40* (1980), 352–356

Authors: G. Müller, K. Norpoth
Approved by the Working Group: 21st February 1986

Documentation for Carcinogenic Substances

Without Biological Exposure Equivalents

Acrylonitrile

Synonyms	cyanoethylene vinyl cyanide propenenitrile
Formula	C_3H_3N $CH_2{=}CH{-}CN$
Molecular weight	53.06
Melting point	–83 °C
Boiling point	–77.30 °C
Vapour pressure at 20 °C	116 hPa
MAK [last established: 1977]	Carcinogenic substance: category IIIA2 of the List of MAK and BAT Values
TRK Value [1993]	3 ml/m^3 7 mg/m^3

1 Metabolism and Kinetics

The discovery of the carcinogenicity of acrylonitrile has caused a great amount of interest in investigations of the metabolism of this substance. As there are hardly any appropriate data from humans, recourse has to be made mainly to animal experimental investigations, which demonstrate the principles of the biotransformation of acrylonitrile.

The available data (17) show that acrylonitrile is taken up at the workplace both by inhalation and percutaneously. Correspondingly it is designated with an "H" notation in the MAK Values List.

1.1 Metabolism

Comparative in vitro investigations with the liver microsomes of rats, mice and humans (2, 38) have provided indications that with respect to the oxidative transformation of acrylonitrile, mice are more different from man than are rats. Most of the experimental investigations on the metabolism of acrylonitrile carried out so far involve the last species.

After inhalation of acrylonitrile, a small amount is exhaled or excreted unchanged in the urine (42). Most of the acrylonitrile is conjugated with glutathione (4, 19). Conclusions as to the underlying enzymatic and non-enzymatic processes can be drawn from the kind of excretion products. The enzyme reactions important in these processes occur in mammals predominantly in the liver. Acrylonitrile is transformed in the animal both enzymatically and also non-enzymatically in the sense of a cyanoethylation (Michael condensation) (3) with the SH-group of glutathione to the corresponding glutathione conjugate (18). This conjugate is the starting point for the series of reactions which finally yield *N*-acetyl-*S*-(2-cyanoethyl)-L-cysteine which is found in the urine. Also functional SH-groups in proteins can be affected by a non-enzymatic cyanoethylation by acrylonitrile which can lead to impairment in enzyme function (37).

Most of the incorporated acrylonitrile is excreted as cyanoethyl-"mercapturic acid" (*N*-acetyl-*S*-(2-cyanoethyl)-L-cysteine). A small amount of the acrylonitrile appears in the urine in the form of other metabolites such as *N*-acetyl-*S*-(2-hydroxyethyl)-L-cysteine, 4-acetyl-3-carboxy-5-cyanotetrahydro-1,4-2*H*-thiazine, cyanoacetic acid and 3-hydroxypropionitrile, which indicate an oxidative metabolism of the substance in the body; the nitrile group is split off and later excreted in urine as thiocyanate (rhodanide). *S*-Carboxymethylcysteine and thiodiglycolic acid are other metabolites without a cyano-group (29). Finally part of the acrylonitrile total internal exposure can penetrate the erythrocytes where its reactions include that with the N-terminal amino group of the valine to form cyanoethylvaline (6).

Fennel et al. (8) report that in the rat 50 %, in the mouse, however, 80 % of the metabolites excreted in urine are formed via the intermediate epoxide (cyanoethylene oxide). This difference is regarded as important for estimation of the risk for man (20).

The metabolism diagram shown in figure 1 summarises the data published to date. The substance is exhaled unchanged to a small extent (7 %) (41). It is bound to glutathione (GSH = γ-L-glutamyl-L-cysteinyl-glycine) non-enzymatically (15, 16) and enzymatically via glutathione transferases (18) and then appears after the cleavage of glutaminic acid (Glu) and glycine (Gly) and N-acetylation (Ac) of cysteine as *N*-acetyl-*S*-(2-cyanoethyl)-L-cysteine (N-Ac-Cys-S-CH_2-CH_2-CN) in the urine (4, 12, 23, 24, 41). The oxidative transformation via mono-oxygenases yields oxirane carbonitrile (glycidonitrile) (13). The enzymatic conjugation of this epoxide to glutathione is the first step in the formation of the urinary metabolites *N*-acetyl-*S*-(2-hydroxyethyl)-L-cysteine (N-Ac-Cys-S-CH_2-CH_2-CN) (4, 23) and 4-acetyl-3-carboxy-5-cyano-tetrahydro-1,4-2*H*-thiazione (12, 24). The enzyme "rhodanese" transforms the cyanide group set free in the production of the hydroxyethyl product into rhodanide (12, 13, 23, 24). Enzymatic hydrolysis of oxirane carbonitrile via "epoxide hydrolase" yields the instable 2,3-dihydroxypropionitrile, which rearranges with the loss of the nitrile group to glycolaldehyde (14, 23, 24). The nitrile group appears again as rhodanide in urine. The oxirane carbonitrile can rearrange spontaneously to the intermediate cyanoacetaldehyde, which is the starting material for formation of both urinary metabolites 3-hydroxypropionitrile (HO-CH_2-CH_2-CN) and cyanoacetic acid (HOOC-CH_2-CN) (25). The epoxide metabolism demonstrated in animal experiments has in the meantime been confirmed in persons exposed to acrylonitrile on the basis of the metabolites found (26).

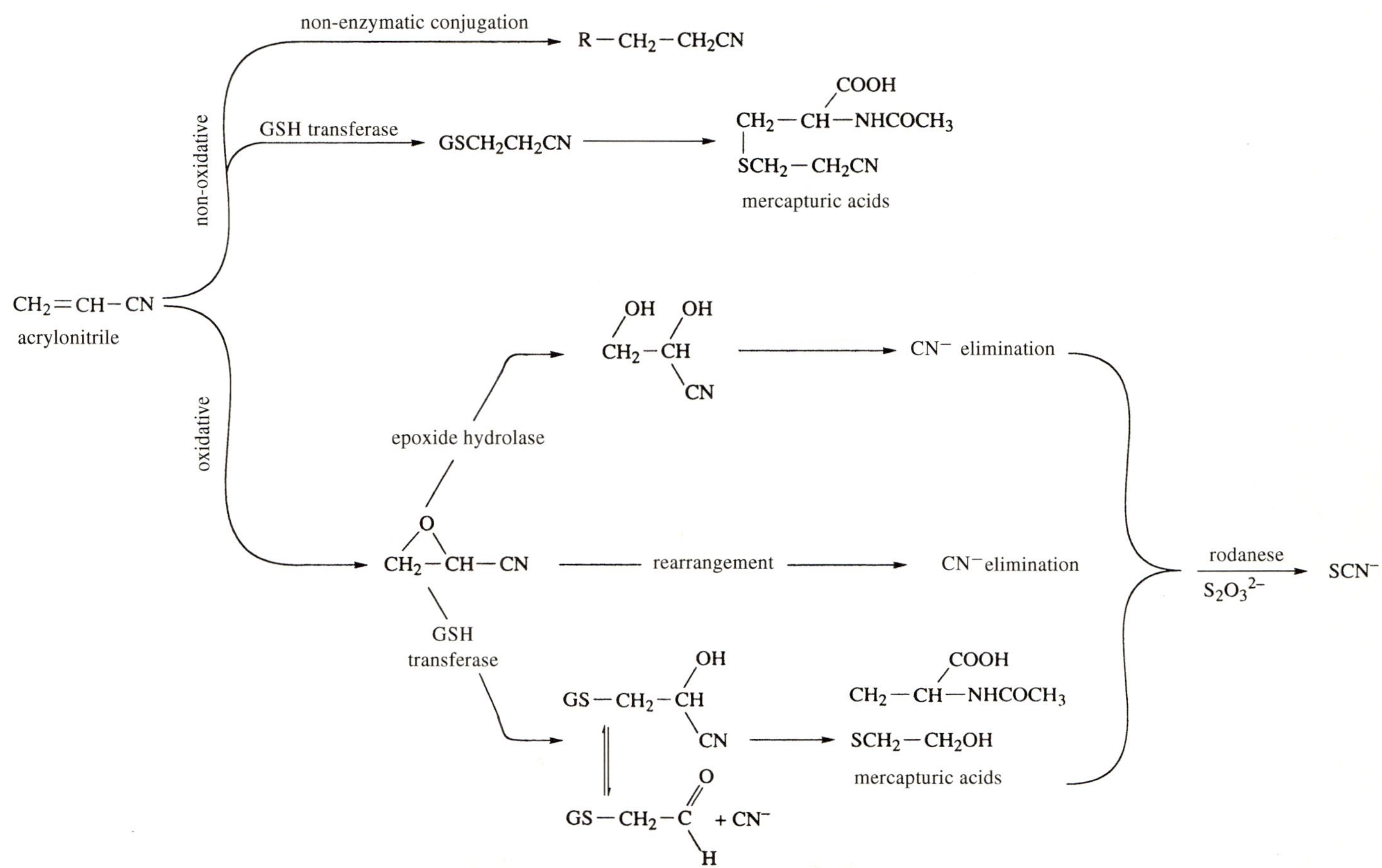

Fig. 1: Relationships between the concentration of acrylonitrile in the ambient air and the urinary excretion of cyanoethylmercapturic acid and acrylonitrile and the cyanoethylvaline concentration in the globin of the erythrocytes in employees with continous exposure to acrylonitrile.

In humans, according to casuistic observations, there appear to be considerable differences in individual disposition which are recognisable in the profile of the acrylonitrile metabolites in urine and in individual differences in the cyanoethylvaline concentration in the globin of erythrocytes (26).

1.2 Kinetics

Investigations after inhalation of the substance by rats and rhesus monkeys have shown that acrylonitrile is retained almost quantitatively from the ambient air (34). After intraperitoneal application to rats only approx. 7 % of the applied acrylonitrile was exhaled unchanged (41). The retained acrylonitrile is metabolised via the routes of metabolism discussed in section 1.1.

There are no kinetic data available for humans.

2 Critical Toxicity

The present MAK value documentation (17) states that acrylonitrile has been shown in animal experiments to have carcinogenic potential. Under inhalation conditions in rats tumours of the mammary gland, Zymbal gland, the forestomach and the brain have been described (35, 28). The evaluation of the biological effect of industrial exposures to acrylonitrile is therefore oriented exclusively to the possibility that the substance causes cancer. The substance is classified in category A2 of section III of the MAK Values List (17).

In addition acrylonitrile has acute toxic effects which are of importance from an occupational-medical point of view in the case of accidents. The most important biological effects are summarised below.

2.1 Carcinogenic Effect

On the basis of biochemical investigations on subcellular fractions and isolated enzymes the hypothesis was put forward by Guengerich et al. (12) that the tumour-producing effect of acrylonitrile is caused by the epoxide (oxirane carbonitrile, glycidonitrile) formed in the organism. Oxirane carbonitrile was a much stronger mutagen than acrylonitrile in S. typhimurium TA100 and TA1535 (11) and in human lymphoblasts (36) and produced more marked damage to SV-40 deoxyribonucleic acids (32).

In vivo (33) and in isolated hepatocytes (9) acrylonitrile and its metabolites have been shown to cause alkylation of DNA but to a much smaller extent than is seen with the structurally related vinyl chloride which has been investigated in similar systems.

2.2 Vegetative Nervous System

After acute intoxication with acrylonitrile symptoms are observed such as also occur after marked stimulation of the parasympathetic system. In animal experiments salivation, vomiting and convulsions were reported (17). In the isolated, perfused heart of a hen under vagal stimulation, addition of acrylonitrile to the perfusate (10 mM) caused an increased release of acetylcholine (34). Occasional reports of workers exposed to acrylonitrile describe increased levels of acetylcholine in the blood (1).

These effects, probably caused by the unmetabolised acrylonitrile itself, are of importance in acute accidental intoxications, not, however, after low-level occupational exposure at concentrations below the currently valid TRK value of 7 mg/m^3.

2.3 Other Organs

In animal experiments an adrenal apoplexy of Waterhouse-Friedrichsen syndrome character is produced after acute high doses of acrylonitrile (45). It has been discussed whether this is connected with a depletion of glutathione in this organ (46). This pattern of damage has no importance for realistic exposure levels at the workplace.

Acute application of high doses of acrylonitrile also causes damage to many other organs. Liver and kidney damage occurs (21) but it is generally only slight (43). In addition, experimental pulmonary oedema (7) and cerebral oedema (5) have been reported occasionally. In the gastrointestinal tract haemorrhage (10) and duodenal ulcers (44) were found. These patterns of damage also occur after high doses in animal experiments.

Early studies of persons exposed to acrylonitrile report the occurrence of liver damage (22, 39, 47, 48). This could not, however, be confirmed in later studies and is probably connected with earlier higher-level exposures, possibly also with exposures to mixtures (42). Such damage is not to be expected if the TRK value for acrylonitrile of 3 ppm is observed.

3 Selection of the Indicators

In line with the concept of exposure and effect, indicators which are related as closely as possible to the effect on the organism should be preferred. For the reasons mentioned above (see section 2.1), determination of metabolites which are formed from the epoxide intermediate (oxirane carbonitrile) should be given preference (29). This prerequisite would theoretically be fulfilled by the determination of thiocyanate (rhodanide) in urine as the product of excretion of the cyanide ion split off as a consequence of epoxidation. There is, however, no data available yet. In addition it should be noted that in smokers the thiocyanate excretion is increased above that in non-smokers. This is also true for the metabolites thiodiglycolic acid and 4-acetyl-3-carboxy-5-cyanotetrahydro-1,4-2*H*-thiazine suggested by the results of animal experiments (29).

The excretion of the main metabolites *S*-cyanoethyl-*N*-acetyl-cysteine (cyanoethylmercapturic acid) and the concentration of acrylonitrile itself in blood or urine can be used as an alternative as parameters of the dose effective upon the organism (29).

Most practical experience published to date for the biological monitoring of exposed persons is available for the excretion of acrylonitrile in urine (19, 40). Further parameters must then be taken into consideration when field experience is available.

4 Recommended Methods

The determination of the concentration of acrylonitrile in urine is carried out by head space gas chromatography. The procedure is described by Houthuijs et al. (19). An earlier method of sample preparation by azeotropic distillation of the urine (42) obviously involves considerable preparation losses.

5 Exposure and Effects

The determination of acrylonitrile excretion is to be regarded as the exposure indicator. Indicators of effect are not available to date.

6 Evaluation of the Biological Exposure Equivalents for Carcinogenic Substances (EKA)

Acrylonitrile, as a carcinogenic substance, is not given a BAT value as at present no biological value can be given that is regarded as safe. The following treatise is intended to outline the current state of the art in occupational medicine for the detection and quantification of individual acrylonitrile exposures (see section VI "Significance and Usage of BAT Values" of the MAK/BAT Values List). When further scientific data are available it should be amended or edited.

To date the data available from biological monitoring of persons exposed to acrylonitrile record the acrylonitrile excretion in urine. The results which were obtained by Houthuijs et al. (19) from a very limited collective of workers, have, however, been confirmed by observations by the working group "Setting of Threshold Limit Values in Biological Material" in other collectives (see table 1).

Since 1965 it has been known that acrylonitrile occurs in tobacco smoke. The detection of acrylonitrile excretion by smokers otherwise not exposed has, however, only been possible since the availability (40) of the improved method of head space gas chromatography (19). The fact that in some of the animal experimental studies on rats no excretion of acrylonitrile in urine could be detected (4, 24) while other authors succeeded in doing this (29) is to be seen in this context.

As early as 1978 Sakurai et al. described the direct determination of acrylonitrile in urine for the evaluation of contact with technical grade acrylonitrile (40). These authors were able to attribute the external acrylonitrile exposure of 0.1 ppm (0,23 mg/ m^3) to an excretion of 3.9 µg acrylonitrile/l urine.

Further investigations on exposed workers are necessary. The data available to date are shown in table 1.

Tab. 1: Relationships between the external concentration of acrylonitrile (ACN) in the ambient air and the urinary excretion of acrylonitrile and cyanoethylmercapturic acid and the cyanoethylvaline concentration in the globin of the erythrocytes in employees continually exposed to acrylonitrile (27)

Persons/ smoking status	Air mg ACN/m^3	Urine µg ACN/l	Urine µg cyano-ethylmer-capturic acid	Blood µg cyano-ethylvaline/l
27/smokers	1)	<10	./.	7 (5–10)
20/non-smokers	1)	<10	./.	2 (<1–4)
	0.3	39[2]	./.	./.
41/smokers }	0.5	85	98	11 (6–15)
16/non-smokers }		(10–130)	(36–178)[3)]	6 (2–10)
39/smokers }	1.0	148	2620	13 (5–20)
17/non-smokers }		(25–210)	(100–6600)[3)]	8 (2–12)
(extrapolated)	7.0 (TRK value)	1000		

./. not measured
1) persons without industrial contact with ACN
2) Houthuijs et al. (19)
3) Müller and Norpoth (31)

The study by Houthuijs et al. (19) on the excretion of acrylonitrile in urine was carried out on 15 exposed workers. During a 7-day shift period and subsequent 2 days rest all urine samples were collected separately. Personal air sampling showed on average for the collective an 8-hour average concentration in air of 0.13 ppm. The urinary concentrations of acrylonitrile typically reached a maximum after the end of the shift and decreased rapidly in the exposure-free interval until the next shift. In a control collective of 41 non-exposed persons from the same company, the urinary excretion of acrylonitrile was significantly dependent on the number of cigarettes smoked; however the acrylonitrile excretion by the exposed persons was always much higher than that of the non-exposed, even during the exposure-free interval.

In addition to the acrylonitrile excretion in the urine, also the urinary excretion of the mercapturic acids cyanoethylmercapturic acid and hydroxyethylmercapturic acid is accessible to analysis (30). First results of measurements of the cyanoethylmercapturic acid excretion are also contained in table 1. A further parameter is the content of cyanoethylvaline in the haemoglobin of the erythrocytes (6). The data (Table 1) are very promising for future investigations, do not yet, however, allow an evaluation of an EKA correlation.

On the basis of the available data (Table 1) the following provisional correlation between the acrylonitrile concentration in the air and in the post-shift urine can be evaluated:

Acrylonitrile concentration	
Air (mg/m^3)	Urine (µg/l)
0.3	40
0.5	80
1.0	150
7.0 (extrapolated)	1000

7 Interpretation of the Data

The interpretation of data on the internal exposure to acrylonitrile must, where the values are low, take into consideration the influence of smoking habits on the results.

The acrylonitrile excretion determined in urine is not of equal value to the cellular exposure measured in the Hb conjugate. The erythrocyte acrylonitrile exposure is to be regarded as a result of the individual metabolic performance, including the individual glutathione transferase (detoxification) potential and monooxygenase (toxification) potential of the organism.

This individual metabolic situation can be assessed by the acrylonitrile conjugation with globin as a marker. By the determination of the erythrocytic cyanoethylvaline conjugates not only the current acrylonitrile exposures but also those up to 3 months previously are recognised. Nevertheless, the database does not yet allow the setting of EKA values.

Investigations in the erythrocytes are not bound to any particular sampling time. On the other hand, the urine investigations provide a reliable estimate of acrylonitrile exposure only if sampling takes place immediately after exposure.

Occupational-medical health surveillance of persons exposed to acrylonitrile should otherwise be carried out in Germany according to the "berufsgenossenschaftlicher Grundsatz" G40 (guideline issued by the Employers' Liability Insurance Association).

8 References

1. Ageeva, T. S.: Tr. Sarat. Med. Inst. *78* (1979), 10–13
2. Appel, K. E., Peter, H., Bolt, M., Bolt, H. M.: Interaction of acrylonitrile with hepatic microsomes of rats and men. Toxicol. Lett. *7* (1981), 335
3. American Cyanamide Company: American Cyanamide Company, Petrochemicals Department, New York, 2.A. (1959)

4. van Bladeren, P. J., Delbressine, L. P. C., Hoogeterp, J. J., Beaumont, A. H. M. G., Beimer, D. D., Seutter-Berlage, F., van der Gen, A.: Formation of mercapturic acids from acrylonitrile, crotononitrile, and cinnamonitrile by direct conjugation and via an intermediate oxidation process. Drug Metab. Disp. *9* (1981), 246–249
5. Brieger, H., Rieders, F., Hodes, W. A.: Acrylonitrile: spectrophotometric determination, acute toxicity, and mechanism of action. Arch. Int. Hyg. Occup. Med. *6* (1952), 128–140
6. Bryant, M. S., Osterman-Golkar, S. M.: Hemoglobin adducts as dosimeters of exposure to DNA-reactive chemicals. CIIT Activities *11* (1991), 1–12
7. Dudley, H. C., Neal, P. A.: Toxicology of acrylonitrile (vinyl cyanide). I. A study of the acute toxicity. J. Ind. Hyg. *24* (1942), 27–36
8. Fennell, T. R., Kedderis, G. L., Sumner, S. C. J.: Urinary metabolites of 1,2,3-acrylonitrile in rats and mice detected by ^{13}C nuclear magnetic resonance spectroscopy. Chem. Res. Toxicol. *4* (1991), 678–687
9. Geiger, L. E., Hogy, L. L., Guengerich, F. P.: Metabolism of acrylonitrile by isolated rat hepatocytes. Cancer Res. *43* (1983), 3080–3087
10. Ghanayem, B. J., Ahmed, A. E.: Acrylonitrile-induced gastrointestinal hemorrhage and the effects of metabolism modulation in rats. Toxicol. Appl. Pharmacol. *68* (1983), 290–296
11. Cerna, M., Kocisova, J., Kodytkova, J., Kopecky, J., Sram, R. J.: Mutagenic activity of oxiranecarbonitrile. In: Gut/Cikrt/Plaa (eds.): Industrial and Environmental Xenobiotics. Springer Verlag, Berlin 1981, pp. 251–254
12. Guengerich, F. P., Geiger, L. E., Hogy, L. L., Wright, P. L.: In vitro metabolism of acrylonitrile to 2-cyanoethylene oxide, reaction with glutathione and irreversible binding to proteins and nucleic acids. Cancer Res. *41* (1981), 4925– 4933
13. Gut, I., Nerudova, J., Kopecky, J., Holecek, V.: Acrylonitrile biotransformation in rats, mice, and chinese hamsters as influenced by the route of administration and by phenobarbital, SKF 525-A, cysteine, dimercaprol or thiosulfate. Arch. Toxicol. *33* (1975), 151–161
14. Gut, I., Kopecky, J., Nerudova, J., Holecek, V.: Pracov. Lek. *28* (1976), 110–118
15. Hashimoto, K., Kanai, R.: Studies on the toxicology of acrylonitrile. Metabolism, mode of action and therapy. J. Ind. Health *3* (1965), 30–46
16. Hashimoto, K., Kanai, R.: Effect of acrylonitrile on sulphydryls and pyruvate metabolism in tissues. Biochem. Pharmacol. *21* (1972), 635–640
17. Henschler, D. (ed.): Gesundheitsschädliche Arbeitsstoffe - Toxikologisch-arbeitsmedizinische Begründung von MAK-Werten, 12th issue. VCH Verlagsgesellschaft, Weinheim 1978
18. Holocek, V., Kopecky, J.: Conjugation of glutathione with acrylonitrile and glycidonitrile. In: Gut/Cikrt/Plaa (eds.): Industrial and Environmental Xenobiotics, Springer Verlag, Berlin 1981, pp. 238–244
19. Houthuijs, D., Remiju, B., Willems, H., Boleij, J., Biersteker, K.: Biological monitoring of acrylonitrile exposure. Am. J. Ind. Med. *3* (1982), 313–320
20. Kedderis, G. L.: The biotransformation of acrylonitrile: Implications of metabolic studies for risk assessment. CIIT Activities *9* (1989), 1–5
21. Knobloch, S., Szendzikowski, S., Czajkowska, T., Krysiak, B.: Med. Pracy *22* (1991), 257–269
22. Kondo, H.: Health care of acrylonitrile workers. Rodo Esei *3* (1969), 21–25
23. Kopecky, J., Gut, I., Nerudova, J., Zachardova, D., Holecek, V.: Two routes of acrylonitrile metabolism. J. Hyg. Epid. Microbiol. Immun. *24* (1980), 356–362
24. Langvardt, P. W., Putzig, C. L., Braun, W. H., Young, J. D.: Identification of the major urinary metabolites of acrylonitrile in the rat. J. Toxicol. Environ. Health *6* (1980), 273–282
25. Lambotte-Vandepaer, M., Duverger-van Bogaert, M., de Meester, C., Rollmann, B., Poncelet, F., Mercier, M.: Identification of two urinary metabolites of rats treated with acrylonitrile; influence of several inhibitors on the mutagenicity of those urines. Toxicol. Lett. *7* (1981), 321–328
26. Lewalter, J.: Empfehlungen zur arbeitsmedizinischen Prävention expositions- und dispositionsbedingter Arbeitsstoff-Beanspruchungen. Verh. Dtsch. Ges. Arbeitsmed. *31* (1991), 135–139
27. Lewalter, J.: unpublished investigations
28. Maltoni, C., Ciliberti, A., di Maio, V.: Carcinogenicity bioassays on rats of acrylonitrile administered by inhalation and by ingestion. G. Ital. Med. Lav. *68* (1977), 401–411
29. Müller, G., Wintgens, J., Tenkamp, Th.: Möglichkeiten der biologischen Überwachung bei Einwirken von Acrylnitril. Verh. Dtsch. Ges. Arbeitsmed. *23* (1983), 399–403

30. Müller, G., Werfet, U., Nebe, J., Kneckt, U., Norpoth, K.: Mercaptursäuren im Harn als Überwachungsparameter bei Expositionen gegenüber Morbenzol und Acrylnitril. Verh. Dtsch. Ges. Arbeitsmed. *32* (1992), 432–435
31. Müller, G., Norpoth, K.: unpublished investigations
32. Peter, H., Schwarz, M., Mathiasch, B., Appel, K. E., Bolt., H. M.: A note on synthesis and reactivity towards DNA of glycidonitrile, the epoxide of acrylonitrile. Carcinogenesis *4* (1983), 235–237
33. Peter, H., Appel, K. E., Berg, R., Bolt. II. M.: Irreversible binding of acrylonitrile to nucleic acids. Xenobiotica *13* (1983), 19–25
34. Peter, H., Bolt, H. M.: Experimental pharmacokinetics and toxicology of acrylonitrile. G. Ital. Med. Lav. *6* (1984), 77–81
35. Quast, J. F., Enriquez, R. M., Wade, C. E., Hamiston, C. G. Schwetz, B. A.: Investigations by the DOW Chemical Co., Midland, MI, USA published in part; cited acc. (16.) Federal Register *43* (1978), 2589
36. Recio, L., Skopek, T. R.: Mutagenicity of acrylonitrile and its metabolite 2-cyanoethylene oxide in human lymphoblasts in vitro. Mutat. Res. *206* (1988), 297–305
37. Riehm, J. P., Scheraga, H. A.: Structural studies of ribonuclease. XX. Acrylonitrile. A reagent for blocking the amino groups of lysine residues in ribonuclease. Biochemistry *5* (1966), 93–99
38. Roberts, A. E., Kedderis, G. L., Turner, M. J., Rickert, D. E., Swenberg, J. A.: Species comparison of acrylonitrile epoxidation by microsomes from mice, rats and humans: relationship to epoxide concentrations in mouse and rat blood. Carcinogenesis *12* (1991), 401–404
39. Sakurai, H., Kusomoto, M.: Epidemiological study of health impairment among acrylonitrile workers. J. of Science and Labour *48* (1972), 273–282
40. Sakurai, H., Onodera, M., Utsonomiya, T., Minakuchi, H., Iwai, H., Matsumura, H.: Health effects of acrylonitrile in acrylic fibre factories. Br. J. Ind. Med. *35* (1978), 219–222
41. Sapota, A., Draminski, W.: The fate of [^{14}C]-acrylonitrile in rats. In: Gut/Cikrt/Plaa (eds.): Industrial and Environmental Xenobiotics, Springer Verlag, Berlin 1981, pp. 231–237
42. Sato, M., Ishizu, S., Momotani, H.: Determination of acrylonitrile, cyanide and thiocyanate in blood and urine. Jap. J. Ind. Health *17* (1975), 99–105
43. Silver, H. E., McComb, D. J., Kovacs, K., Szabo, S.: Limited hepatotoxic potential of acrylonitrile in rats. Toxicol. Appl. Pharmacol. *64* (1982), 131–139
44. Silver, E. H., Galagher, G. T., Szabo, S.: Potentiation of acrylonitrile-induced duodenal ulcer by treatment with aroclor-1254 or phenobarbital. Fed. Proc. *41* (1982), 926
45. Szabo, S., Reynolds, E. S., Kovacs, K.: Animal model of human disease. Waterhouse-Friderichsen syndrome. Animal model: Acrylonitrile-induced adrenal apoplexy. Am. J. Pathol. *82* (1976), 653–656
46. Szabo, S., Bailey, K. A., Boor, P. J., Jaeger, R. J.: Acrylonitrile and tissue glutathione: differential effect of acute and chronic interactions. Biochem. Biophys. Res. Commun. *79* (1977), 32–37
47. Wilson, R. H.: Health hazards encountered in the manufacture of synthetic rubber. J. Am. Med. Ass. *124* (1944), 701–702
48. Wilson, R. H., Hough, G. V., McCormick, W. E.: Medical problems encountered in the manufacture of American-made rubber. Ind. Med. and Surg. *17* (1948), 199–200

Authors: H. M. Bolt, J. Lewalter
Approved by the Working Group: 27th May 1993

4-Aminodiphenyl

Synonyms	p-Aminodiphenyl
Formula	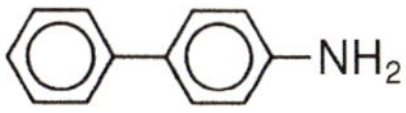
Molecular weight	169.23
Melting point	53-54 °C
Boiling point	302 °C
MAK [last established: 1970]	Carcinogenic substance: category IIIA1 of the List of MAK and BAT Values

Until 1954 4-aminodiphenyl (4-ADP) was used, particularly in the American rubber industry, as a stabilising agent (44). Occupational medical investigations in 1954 showed among 171 exposed persons 19 tumours of the efferent urinary passages and in 1960 53 tumours among 315 exposed persons (13, 25, 43). Today 4-ADP only occurs in small amounts as a by-product, for example in the production and processing of certain amino and nitro aromatics (26). It is contained in traces in tobacco smoke (7, 46, 52, 54, 61). Furthermore, the products of technical combustion processes (diesel soot) include 4-nitrodiphenyl (4-NDP) (10, 18, 41, 62, 63, 65). 4-ADP is listed in the List of MAK and BAT Values in category IIIA1 (11). The most significant effect of exposure to 4-ADP is the formation of tumours of the efferent urinary passages with latency periods of up to 30 years (8, 9, 12, 12a, 14, 44, 58). Acute toxic effects of 4-ADP are not known in humans, neither after inhalation nor after oral or cutaneous exposure (2). In animal experiments (with dogs) formation of methaemoglobin can be induced with 4-ADP (55, 56).

1 Metabolism and Kinetics

4-Aminodiphenyl is absorbed rapidly via the skin and inhalation and excreted within 24 to 48 hours in the form of conjugates of the metabolites, predominantly via the kidneys (7, 17, 54). Low doses of 4-ADP are in general acetylated immediately and thus can be eliminated renally (26, 64). With increasing 4-ADP uptake the action of mixed function oxidases leads to significant increases in particular in hydroxylated metabolites (1, 17, 23, 40). The clinically important *N*-oxidation is more pronounced in "slow acetylators" than in "fast acetylators"; it leads to the formation of *N*-hydroxy-4-ADP (7, 19, 32, 45, 56). The *N*-hydroxy-4-ADP formed in the endoplasmatic reticulum of the liver is stabilised by conjugation with *N*-glucuronide; the conjugate can be cleaved to yield *N*-hydroxy-4-ADP in the target organ (bladder) by β-glucuronidases (21, 27, 53).

In the erythrocytes *N*-hydroxy-4-ADP is oxidised to 4-nitrosodiphenyl. Part of the 4-nitrosodiphenyl formed in the erythrocytes binds covalently to the SH group of haemoglobin or of glutathione forming sulphinic acid amides (6, 7, 17, 20).

The electrophilic activation of the inert *N*-hydroxy-4-acetaminodiphenyl (*N*-hydroxy-4-AADP) takes place in several metabolic steps, for example by conjugation with sulphuric acid, deacetylation to *N*-hydroxy-4-ADP or intramolecular N,O-acyl transfer to *N*-acetoxy-4-ADP (3). In the acidic urine of the bladder (pH: 5-6) the *N*-acetoxy-4-ADP or the *N*-hydroxy-4-ADP which is instable in free form can decompose into the electrophilic diphenylnitrenium ion or into other electrophilic derivatives (27, 28, 53). These bind either directly to macromolecules of the bladder epithelium or are detoxified by conjugation and excreted (7, 27, 37, 42, 51, 53). The individual reaction steps are represented in figure 1 (7, 27, 29).

The relatively long biological half-life of the erythrocytic 4-ADP adducts leads to accumulation, which makes a correlation with the 4-ADP adduct dose in other compartments difficult (20). For this reason the determination of the haemoglobin adducts can only provide general monitoring of the internal dose of 4-ADP, not specific monitoring of the target organ (7).

Variability of the Effect of 4-ADP

Adduct formation with proteins such as, e.g., haemoglobin (Hb) (34) and thus probably also with nucleic acids is largely avoided if the incorporated 4-aminodiphenyl is completely acetylated and thus eliminated quickly (via the kidneys) (33). Therefore the acetylator status has an important influence on the level of adduct formation which in general is particularly pronounced in "slow acetylators" (34, 48).

In 1979 attention was drawn for the first time to the association between the acetylator status and the development of bladder carcinomas (34, 36, 38, 47). In epidemiological studies it was discovered that fast acetylators were represented to a markedly lesser extent among patients with tumours of the efferent urinary passages than were slow acetylators despite comparable exposures to carcinogenic amino aromatics (19, 22, 36, 38, 47, 48). Hansen et al. found that of 105 persons with a bladder tumour 61.9 % were slow acetylators compared to only 42.9 % of patients without bladder carcinomas (22). With occupationally-induced bladder tumours the ratio was even higher 70 : 30 (22). Observations by Lewalter in collectives with complete occupational anamneses show a still more significant ratio (100 : 0) (32). Not only the amount of the 4-ADP taken up but also the interindividual variance of the metabolic pathways is decisive for the kind and extent of the 4-ADP protein adducts and 4-ADP-DNA adducts (39) and thus for possible genotoxic effects formed in the individual (3, 16, 49, 60).

In general detectable or conspicuously elevated 4-ADP haemoglobin adduct levels are only observed in slow acetylators (32).

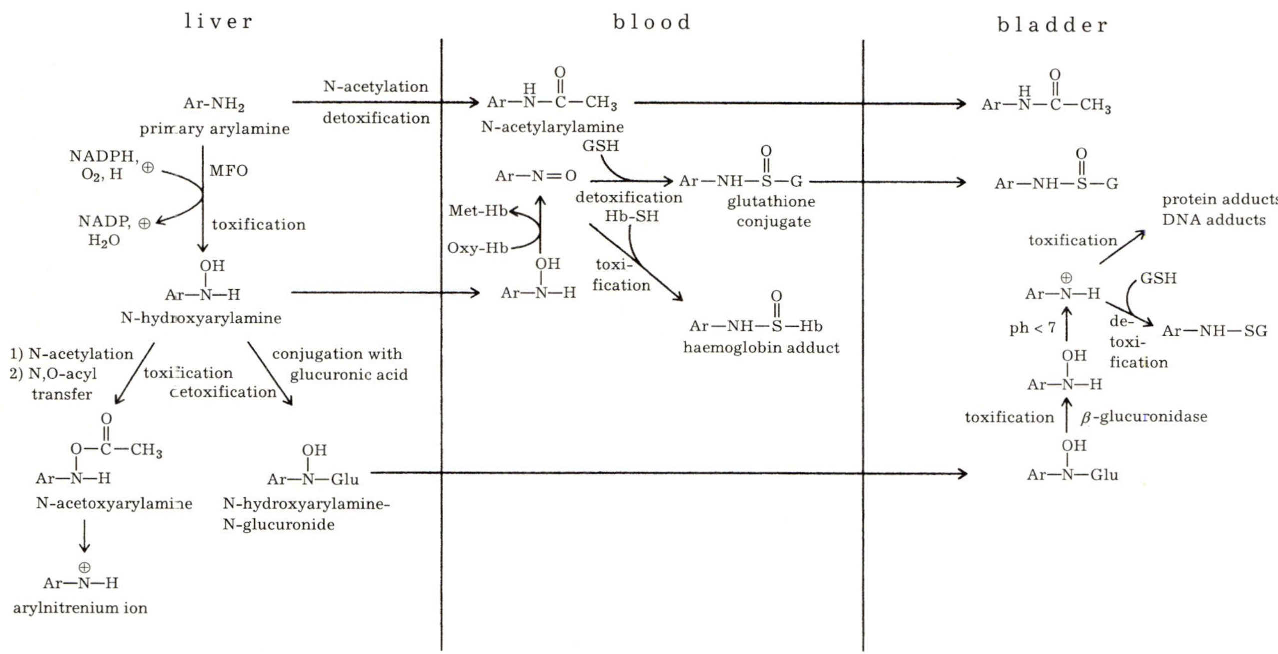

Fig. 1: Metabolism of primary amino aromatics ($Ar\text{-}NH_2$): formation and distribution of the metabolites (7, 27, 29); MFO = mixed function oxygenases, GSH = glutathione

2 Critical Toxicity

Presently available data have proved the carcinogenic effect of 4-ADP in humans.

Unlike the biological monitoring of aniline for which the BAT value is oriented on the Met-Hb level (24), the biological monitoring of persons exposed to 4-ADP must be focused on the minimising or avoidance of the macromolecular binding of 4-ADP.

3 Exposure and Effects

With 4-ADP every exposure which can lead to inhalation or dermal absorption is to be regarded as external exposure. The internal exposure is the biologically relevant concentration of the reactive intermediate in the target organ for which the carcinogenic effect is known, above all in the bladder.

A parameter of internal exposure is the 4-ADP concentration in body fluids such as blood, plasma and urine.

The 4-ADP adducts in the erythrocytes (Hb adducts) or leucocytes (DNA adducts) are understood as parameters of the internal effect of 4-ADP.

With the monitoring of the incidence of DNA strand breaks in leucocytes of persons exposed to 4-ADP a further parameter of effect is available, albeit unspecific and mainly suitable for the evaluation of acute 4-ADP intoxications.

4 Selection of the Indicators

4.1 4-Aminodiphenyl in Urine

It has been possible for some time now to measure the 4-ADP excretion (as conjugates) in the urine of persons exposed occupationally to 4-ADP (35, 50). Nevertheless, a measured correlation between the inhaled 4-ADP concentrations and the urinary excretion of the 4-ADP conjugates is still not available. The reasons for this include the frequently predominantly dermal 4-ADP exposures which cannot be evaluated quantitatively.

In practice, therefore, the excretion of 4-ADP in the urine of persons exposed to 4-ADP is assessed on the basis of occupational medical experience (31).

4.2 4-Aminodiphenyl Protein Adducts

According to the investigations of various research groups, glutathione conjugates are formed in the reaction of the metabolite 4-nitrosodiphenyl with the thiol group of glutathione (15) and corresponding adducts with the thiol groups of proteins (7). In particular the 4-ADP adducts with serum albumin (HSA) can be used for biological monitoring (7).

In order to restrict the amount of work required to detect these 4-ADP adducts it is recommended to test for 4-ADP adduct formation by HSA screening and only when these results are conspicuous or positive to quantify the Hb adducts. For the reasons already mentioned in section 4.1, there are no investigations on the correlation between defined Hb and HSA adducts and the level of the external exposure to 4-ADP.

4.3 DNA Adducts

Within the DNA the position C8 in guanine is the preferred target for the binding of active metabolites of 4-ADP. The reaction with *N*-hydroxy-4-ADP yields *N*-(deoxyguanosine-8-yl)-4-ADP (30). Biological monitoring of 4-ADP-DNA adducts in leucocytes is possible in principle. To what extent the detection of DNA adducts can be correlated with an effect on the DNA requires, however, further investigation (16, 49, 60). A genotoxic effect can also be detected by determining the incidence of DNA strand breaks (5).

5 Methodology

The analysis of 4-ADP takes place by high pressure liquid chromatography or (after derivatisation) capillary gas chromatography with an ECD detector or by GC/MS. HSA adducts are determined by isoelectric focusing. DNA adducts in the leucocytes are analysed with the GC/MS technique. The methods of analysis are described in the literature (7, 27, 59).

Determination of the incidence of DNA strand breaks in the leucocytes makes use of the method of alkaline elution (57).

Evaluation of the 4-ADP Content of Haemoglobin (Hb) Adducts and DNA Adducts

The 4-ADP content of the Hb adducts (μg 4-ADP_{Hb}) and the DNA adducts in the leucocytes (μg 4-ADP_{DNA}) are calculated, taking into account possible losses during preparation by use of the corresponding packed cell volumes (PCV) expressing the PCV of the erythrocytes or leucocytes in terms of the PCV of whole blood according to:

$$c_{Hb/DNA}\left(\frac{\mu g\ 4\text{-ADP}}{1\ \text{blood}}\right) = \frac{PCV_{blood}}{PCV_{suspension}} \cdot c_{suspension}\left(\frac{\mu g\ 4\text{-ADP}}{1\ \text{suspension}}\right)$$

$c_{Hb/DNA}$ = amount of 4-aminodiphenyl bound to haemoglobin (erythrocytes) or to DNA (leucocytes) per volume of blood

$c_{suspension}$ = amount of 4-aminodiphenyl measured in the erythrocyte or leucocyte suspension

PCV_{blood} = packed cell volume of whole blood

$PCV_{suspension}$ = packed cell volume of the erythrocyte or leucocyte suspension

6 Evaluation of the Biological Exposure Equivalents for Carcinogenic Substances (EKA)

6.1 4-Aminodiphenyl in Urine

Despite the monitoring of 4-ADP excretion in the urine of persons exposed to amino aromatics in industry, there are no published investigations of its dependence upon inhaled 4-ADP concentrations. Urinary excretion of 4-ADP detectable after contact with amino aromatics can be most probably explained by dermal absorption. Interindividual differences in the 4-ADP excretion rate are due to individual differences in the capability to acetylate 4-ADP (48).

In practice, for the occupational medical monitoring of exposure to 4-ADP "4-ADP monitoring values" of the order of magnitude of 100 ng 4-ADP per g creatinine have been named, however without scientific justification (4, 31).

6.2 4-Aminodiphenyl in Blood Cells

In an investigation of 29 smokers (on average 20 cigarettes/day) the 4-ADP content of the 4-ADP-Hb adducts was found to be 24.4 ng per l blood (7). The 4-ADP dose taken up daily amounted to approx. 100 ng in the main stream smoke and approx. 3000 ng 4-ADP in the side stream smoke.

The information available to the Commission at present indicates that 4-ADP exposures which cause urinary excretion levels below 100 ng per g creatinine and 4-ADP-Hb adduct levels below 100 ng per l blood do not produce detectable levels of DNA adducts in the leucocytes. In practice, the biological monitoring of persons exposed to 4-ADP can be based on these values. The establishing of correlations with the exposure is, with the database available at present, not possible for the reasons mentioned above.

7 Interpretation of the Data

For purposes of prevention, the 4-ADP concentrations measured in urine are not equal in value with the effect values measured in blood cells.

As a routine procedure, the investigation of the erythrocytic 4-ADP-Hb adducts is to be recommended as it determines not only current exposures but also any exposure from one week to two months previously (7).

Investigations in erythrocytes are not bound to any particular sampling time (31). On the other hand, urine investigations only provide reliable results if the sampling is carried out immediately after exposure (31).

8 References

1. Bayraktar, N., Kajbaf, M., Jatoe, S. D., Gorrod, J. W.: The oxidation of isomeric amino and acetaminodiphenyl by rat hepatic microsomal preparations. Arch. Toxicol. *60* (1987), 91– 92
2. Beard, R. R., Nee, J. T.: Aromatic nitro and amino compounds. In: Clayton, G. D., Clayton, F. E. (eds.): Patty's Industrial Hygiene and Toxicology, 3rd rev. ed. Vol. 2A, Toxicology. Wiley & Sons, New York 1981, pp. 2413–2489
3. Beland, F. A., Kadlubar, F. F.: Formation and Persistence of arylamine DNA adducts in vivo, Environ. Health Perspect. *62* (1985), 19–30
4. Berufsgenossenschaft der chemischen Industrie (ed.): Anerkannte Analysenverfahren zur Feststellung der Konzentration krebserzeugender Arbeitsstoffe in der Luft in Arbeitsbereichen. ZHI/120. Heymann, Köln 1981, pp. 47–56
5. Blome, H., Thielen, R. G., Henning, M.: Untersuchungen über die Belastung von Kokereiarbeitern durch 2-Naphthylamin. Staub-Reinhalt. Luft *43* (1983), 459–461
6. Bonse, G., Metzler, M.: Biotransformation organischer Fremdsubstanzen. Thieme, Stuttgart 1978, pp. 101–102
7. Bryant, M. S., Skipper, P. L., Tannenbaum, S. R., Maclure, M.: Hemoglobin adducts of 4-aminodiphenyl in smokers and non-smokers. Cancer Research *47* (1987), 602–608
8. Case, A. M.: Some environmental carcinogens. Proc. R. SOG. Med. *62* (1969), 1061–1066
9. Couney, A. H.: Induction of microsomal enzymes by foreign chemicals and carcinogenesis by polycyclic aromatic hydrocarbons. In: G. H. A. Clowes memorial lecture. Cancer Res. *42* (1982), 4875–4917
10. Deichmann, W. B., Radomski, J. L., Anderson, W. A. D., Coplan, M. M., Woods, F. M.: The carcinogenic action of p-aminobiphenyl in the dog. Ind. Med. Surg. *27* (1958), 25–26
11. Deutsche Forschungsgemeinschaft (ed.): Protokoll der Austauschkonferenz 22. 11. 1957. Arbeitsgruppe: Blasenkrebs, Bonn 1957
12. Deutsche Forschungsgemeinschaft in: Horbach, L., Loskant, H. (eds.): Berufskrebsstudie. Forschungsbericht. Boldt, Boppard 1981

12a. Doerjer, G., Bedell, M. A., Oesch, F.: DNA adducts and their biological relevance. In: Obe, E. (ed.): Mutations in man. Springer, Berlin 1984, pp. 20–34

13. Ehrlicher, H.: Krebs als Berufskrankheit. Arbeitsmed. Sozialmed. Arbeitshyg. *6* (1971), 221–225
14. Eyer, P., Lierheimer, E.: Biotransformation of nitrosobenzene in the red cell and the role of glutathione. Xenobiotica *10* (1980), 517–526
15. Eyer, P.: The red cell as a sensitive target for activated toxic arylamines. Arch. Toxicol. Suppl. *6* (1983), 3–12
16. Flammang, T. J., Kadlubar, F. F.: Acetyl-CoA-dependent, cytosol-catalyzed binding of carcinogenic N-hydroxy-arylamines to DNA. In: Boobies, A. R., Caldwell, J., DeMatteis, F., Elcombe, C. R., Microsomes and Drug Oxidations. Taylor and Francis, London 1985, pp. 190–197
17. Gibson, T. L.: Sources of direct-acting nitroarene mutagens in airborne particulate matter. Mutat. Res. *122* (1983), 115–121
18. Glowinski, J. B., Radtke, H. E., Weber, W. W.: Genetic variation in N-acetylation of carcinogenic arylamines by human and rabbit liver. Mol. Pharmacol. *14* (1978), 940–949
19. Green, L. C., Skipper, D. L., Turesky, R. J., Bryant, M. S., Tannenbaum, S. R.: In vivo dosimetry of 4-aminobiphenyl in rats via a cysteine adduct in hemoglobin. Cancer Res. *44* (1984), 4254–4259
20. Gupta, R. C., Dighe, N. R.: Formation and removal of DNA adducts in rat liver treated with N-hydroxy derivatives of 2-acetylaminofluorence, 4-acetylaminobiphenyl and 2-acetylaminophenanthrene. Carcinogenesis *5* (1984), 343–349
21. Hanna, P. E., Banks, R. B.: Arylhydroxylamines and arylhydroxamic acids: conjugation reactions. In: Anders, M. W. (ed.): Bioactivation of foreign compounds. Academic, Orlando 1985, pp. 375–402
22. Hauptverband der gewerblichen Berufsgenossenschaften (ed.): Berufsgenossenschaftliche Grundsätze für Arbeitsmedizinische Vorsorgeuntersuchungen. Ergänzung 1985. Gentner, Stuttgart 1985
23. Hendry, J. A., Matthews, J. J., Walpole, A. L., Williams, M. H. C.: Nature *175* (1955), 1131
24. Henschler, D., Lehnert, G. (eds.): Biologische Arbeitsstoff-Toleranz-Werte (BAT-Werte). Arbeitsmedizinisch-toxikologische Begründungen. Bd. 1, 3. Lfg., VCH Verlagsgesellschaft, Weinheim 1986

25. International Agency for Research of Cancer: IARC Monographs on the Evaluation of the Carcinogenic Risk of Chemicals to Humans. Vol. 1 : Some inorganic substances, chlorinated hydrocarbons, aromatic amines, N-nitroso compounds, and natural products. IARC, Lyon 1972, p. 74
26. Jatoe, S. D., Gorrod, J. W.: The in vitro/in vivo comparative metabolism of 4-aminobiphenyl using isolated hepatocytes. Arch. Toxicol. *60* (1987), 65–68
27. Kadlubar, F. F., Miller, J. A., Miller, E. C.: Hepatic microsomal N-glucuronidation and nucleic acid binding of N-hydroxy arylamines in relation to urinary bladder carcinogenesis. Cancer Res. *37* (1977), 805–814
28. Kasper, Ch. B., Henton, D.: Glucuronidation. In: Jakoby, W. B. (ed.): Enzymatic basis of detoxication. Vol. II. Academic Press, New York 1980, pp. 9–13
29. King, C. M., Allaben, W. T.: Arylhydroxyamic acid acyltransferase. In: Jakoby, W. B. (ed.): Enzymatic basis of detoxication. Vol. II. Academic Press, New York 1980, pp. 187–197
30. Kriek, E., Westra, J. G.: Metabolic activation of aromatic amines and amides and interactions with nucleic acids. In: Grover, P. L. (ed.): Chemical carcinogens and DNA. Vol. II. CRP, Boca Raton 1979, pp. 1–28
31. Lewalter, J.: Personal experiences during the biological monitoring of persons exposed to amino aromatics. 1978–1987
32. Lewalter, J.: Unpublished observations, 1987
33. Lewalter, J.: Neue methodische Ansätze in der biologischen Überwachung Arbeitsstoff-Exponierter. In: Verhandlungen der Deutschen Gesellschaft für Arbeitsmedizin. 25. Jahrestagung. Gentner, Stuttgart 1985, pp. 151–166
34. Lewalter, J., Korallus, U.: Blood protein conjugates and acetylation of aromatic amines: New findings on biological monitoring. Int. Arch. Occup. Environ. Health *56* (1985), 179–196
35. Linch, A. L.: Biological monitoring for industrial chemical exposure control. CRC, Cleveland 1974
36. Lower jr., G. M.: N-Acetyltransferase phenotype and risk in industrial urinary bladder cancer: approaches to high risk groups. In: Deichmann (ed.): Toxicology and Occupational Medicine. Elsevier, Holland 1979, pp. 209–219
37. Lower jr., G. M.: Arylamines and bladder cancer causality. Application of conceptual and operational criteria. Clin. Pharmacol. Ther. *34* (1983), 129–135
38. Lower jr., G. M., Nilsson, T., Nelson, C. E., Wolf, H., Gamsky, T. E., Bryan, G. T.: N-Acetyltransferase phenotype and risk in urinary bladder cancer: Approaches in molecular epidemiology. Environ. Health Perspect. *29* (1979), 71–79
39. Lu, L.-J. W., Disher, R. M., Randerath, K.: Differences in the covalent binding of benzo(a)pyrene, safrole, l'-hydroxy safrole, and 4-aminobiphenyl to DNA of pregnant and non-pregnant mice. Cancer Letters *31* (1986), 43–52
40. Matthews, J. J., Walpole, A. L.: Br. J. Cancer *10* (1956), 539
41. McCoy, E. C.: Role of metabolism on the mutagenicity of nitroarenes. In: Greim, H., Jung, R., Kramer, M., Marquardt, H., Oesch, F. (eds.): Biochemical Basis of Chemical Carcinogens. Raven, New York 1984, pp. 57–67
42. McQueen, Ch. A., Way, B. M., Williams, G. M.: Biotransformation of aromatic amines to DNA-damaging products by urinary bladder organ cultures. Carcinogenesis *8* (1987), 401–404
43. Melick, W. F., Naryka, J. J., Kelly, R. E.: Bladder cancer due to exposure to para-aminobiphenyl: A 17-year follow-up. J. Ural. *106* (1971), 220–226
44. Melick, W. F., Escue, H. M., Naryka, J. J., Mezera, R. A., Wheeler, E. R.: The first reported cases of human bladder tumours due to a new carcinogen-xenylamine. J. Ural. *74* (1955), 760–766
45. Miller, J. A., Wyatt, C. S., Miller, E. C., Hartmann, H. A.: The N-hydroxylation of 4-acetaminobiphenyl by the rat and dog and the strong carcinogenicity of N-hydroxy-4-acetylaminobiphenyl in the rat. Cancer Res. *21* (1961), 1465–1473
46. Mommsen, S., Aagaard, J.: Tobacco as a risk factor in bladder cancer. Carcinogenesis *4* (1983), 335–338
47. Mommsen, S., Aagaard, J.: Susceptibility in urinary bladder cancer: Acetyltransferase phenotypes and related risk factors. Cancer Letters *32* (1986), 199–205
48. Mommsen, S., Barford, N. M., Aagaard, J.: N-Acetyltransferase phenotypes in the urinary bladder carcinogenesis of a low-risk population. Carcinogenesis *6* (1985), 199–201
49. Neumann, H.-G.: Concepts for assessing the internal dose of chemicals in vivo. In: Fowler, B. A. (ed.): Mechanism of Cell Injury: Implications for Human Health. Wiley and Sons, New York 1987, pp. 241–254

50. Nony, C. R., Bowman, M. C.: Carcinogens and analogs: trace analysis of thirteen compounds in admixtures in waste water and human urine. Int. J. Environ. Anal. Chem. *5* (1978), 203
51. Oglesky, L. A., Flammang, T. J., Tullis, D. L., Kadlubar, F. F.: Rapid absorption, distribution and excretion of carcinogenic N-hydroxyarylamines after direct urethral instillation into the rat urinary bladder. Carcinogenesis *2* (1981), 15–20
52. Patrianakos, C., Hoffmann, D., J.: Chemical studies on tobacco smoke LXIV. On the analysis of aromatic amines in cigarette smoke. Anal. Toxicol. *3* (1979), 150–154
53. Poupko, J. M., Hearn, W. L., Radomski, J. L.: N-Glucuronidation of N-hydroxy aromatic amines. A mechanism for their transport and bladder-specific carcinogenicity. Toxicol. Appl. Pharmacol. *50* (1979), 479–484
54. Radomski, J. L.: The primary aromatic amines: their biological properties and structure-activity relationships. Ann. Rev. Pharmacol. Toxicol. *19* (1979), 129–157
55. Radomski, J. L., Brill, E.: Bladder cancer induction by aromatic amines: Roles of N-hydroxy metabolites. Science *167* (1970), 992–993
56. Radomski, J. L., Conzelmann jr., G. L., Rey, A. A., Brill, E.: N-Oxidation of certain aromatic amines, acetamides, and nitro compounds by monkeys and dogs. J. Natl. Cancer Inst. *50* (1973), 989–995
57. Skare, J. A., Schretel, K. R.: Alkaline elution of rat testicular DNA: Detection of DNA strand breaks after in vivo treatment with chemical mutagens. Mutat. Res. *130* (1984), 283–294
58. Steinborn, J.: Chemische Induktion des Blasenkarzinoms. Zentralbl. Arb. Med. *34* (1984), 258–263
59. Stillwell, W. G., Bryant, M. S., Wishnok, J. S.: GC/MS Analysis of biologically important aromatic amines. Application to human dosimetry. Biomed. Environ. Mass. Spectrom. *14* (1987), 221–227
60. Swenberg, J. A., Richardson, F. C., Boucheron, J. A., Dyroff, M. C.: Relationships between DNA adduct formation and carcinogenesis. Environ. Health Perspect. *62* (1982), 177–183
61. Tannenbaum, S. R., Bryant, M. S., Skipper, P. L., Maclure, M.: Hemoglobin adducts of tobacco related aromatic amines. Banbury Rep. *23* (1986), 63–75
62. Tomkins, B. A., Brazell, R. S., Roth, M. E., Ostrum, V. H.: Isolation of mononitrated polycyclic aromatic hydrocarbons in particulate matter by liquid chromatography and determination by gas chromatography with the thermal energy analyzer. Anal. Chem. *56* (1984), 781–786
63. Uehleke, H., Nerstel, K.: Hydroxylamin- und Nitrosobiphenyl: Biologische Oxidationsprodukte von 4-Aminobiphenyl und Zwischenprodukte der Reduktion von 4-Nitro-biphenyl. Naunyn-Schmiedebergs Arch. Pharmacol. Exp. Path. *257* (1967), 151–171
64. Weber, W. W., Hein, D. W.: N-Acetylation pharmacogenetics. Pharmacol. Rev. *37* (1985), 25–79
65. Wei, E. T., Shu, H. P.: Nitrosoaromatic carcinogens in diesel soot: a review of the laboratory findings. Am. J. Public Health *73* (1983), 1085–1088

Author: J. Lewalter

Approved by the Working Group: 19th November 1987

4,4´-Diaminodiphenylmethane

Synonyms	p,p´-Diaminodiphenylmethane 4,4´-Methylenedianiline MDA
Formula	$C_{13}H_{14}N_2$ 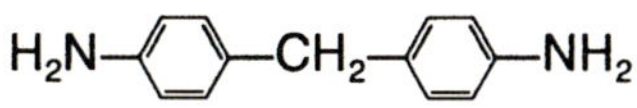
Molecular weight	198.27
Melting point	92 °C
Boiling point	398.5 °C
Vapour pressure at 20 °C	0.06 hPa
MAK [last established: 1987]	Carcinogenic substance: category IIIA2 of the List of MAK and BAT Values
TRK Value [1993]	0.1 mg/m^3

4,4´-Methylenedianiline (MDA) has been produced industrially from aniline and formaldehyde since 1920. Pure MDA consists of pale yellow crystals with a slight amine-like smell. Technical grade MDA contains up to 5 % of the 2,2´-isomer and 2,4´-isomer of the diamine. In general MDA is processed to MDI (4,4´-methylenediphenyl-diisocyanate). Up to 10 % of the MDA is used directly as a cross-linker in epoxy resins, polyurethanes and other polymers as well as in the production of rubber chemicals (4, 12, 15, 19).

1 Metabolism and Kinetics

Details of the metabolism and toxicology of MDA have already been published by the DFG Commission for the Investigation of Health Hazards of Chemical Compounds in the Work Area (22).

Like other aminoaromatics (see 4-aminodiphenyl, aniline, 2-naphthylamine, benzidine) 4,4´-diaminodiphenylmethane is transformed into metabolites which form haemoglobin adducts in erythrocytes (see Tab. 1).

Tab. 1: Haemoglobin adduct level (µg/l) in persons occupationally exposed to high levels of MDA (E. Bailey, unpublished results 1992 (2))

Person	Haemoglobin adduct level (µg/l blood) MDA	N-Acetyl-MDA	Relationship between N-acetyl-MDA and free MDA (mol/mol)	Acetylator phenotype
A	1807	3637	1.7	i
B	414	N	./.	s
C	409	1246	2.5	f
D	536	697	1.1	s
E	639	N	./.	s
F	191	869	5.9	f
G	356	1003	2.3	f
H	242	404	1.4	s
I	639	1064	1.4	s
J	261	1466	4.6	f
K	955	1780	1.5	i
L	463	1785	3.2	f
N = not detectable			Average: 2.6	f : s : i = 42 : 42 : 16 f: fast, s: slow, i: indifferent

2 Critical Toxicity

2.1 Liver Toxicity

MDA feeding experiments with male beagle dogs led, in addition to weight loss, to disturbances in liver metabolism with significantly increased activity of the liver enzymes (13, 42). Oral or dermal exposure or inhalation of MDA on one or a few occasions has also led in humans to disturbances in liver metabolism. In the accident known as “Epping Jaundice” 84 persons contracted hepatitis with markedly increased SGOT, SGPT and bilirubin values after eating bread contaminated with MDA; the bread contained up to 3 mg MDA/kg flour. In one patient haematuria was also observed (25, 26, 39).

Further cases of hepatitis-like diseases have been reported in persons after contact with epoxide resins containing MDA (3, 4, 27, 32).

2.2 Genotoxicity and Carcinogenicity

4,4′-Diaminodiphenylmethane has been classified by the DFG Commission for the Investigation of Health Hazards of Chemical Compounds in the Work Area in category IIIA2. Details can be found in the appropriate documentation (22).

In the Ames test MDA was mutagenic after metabolic activation (36, 37, 43, 44, 52).

In an epidemiological study of 179 MDA workers, three tumours of the efferent urinary passages were observed instead of the expected 0.88 cases (15, 16). This could not be confirmed in two further studies on employees exposed to MDA in a chemical factory and power plant (34, 40, 51).

2.3 Other Effects

The ability of MDA to induce skin sensitisation in humans has been repeatedly reported. The allergen ultimately responsible is unknown. It is possibly a photo-allergen (28, 31). In individual cases impairment of the visual ability has also been described after exposure to MDA. In all cases there was massive exposure to MDA which, however, was not quantified (11, 18, 27, 28).

3 Exposure and Effects

In the case of MDA every exposure is to be regarded as external exposure which can lead to inhalation or percutaneous absorption. The internal exposure is the biologically relevant concentration of the reactive metabolite in the target organ, above all in the liver and in the efferent urinary passages (5, 6, 18). The internal exposure parameters are the MDA concentrations in body fluids such as blood, plasma and urine which correlate with the internal exposure under the specific pharmacokinetic conditions. MDA adducts in erythrocytes (Hb adducts) or leucocytes (DNA adducts) as well as changed activities of liver enzymes are understood as parameters of effect for MDA.

4 Selection of the Indicators

The possible internal exposure and effect after contact with MDA can be determined via analysis of the substances in urine, blood and plasma, as well as via protein and DNA adducts in the blood cells (7, 20).

4.1 MDA in Urine

For some time now, MDA exposure has been measured via the excretion of free and conjugated MDA in urine. Despite extensive studies, exact correlations between inhaled MDA concentrations and the urinary excretion of MDA are not yet available (5, 11, 27, 28). This is not least due to the often predominant percutaneous absorption which cannot be quantitatively determined in the total exposure to MDA. In occupational-medical practice the excretion of MDA after contact with MDA is often judged on the basis of so-called "empirical values" (23, 29, 47).

4.2 MDA Protein Adducts

According to investigations by various research groups, a metabolite of MDA in which an amino group was oxidised to the nitroso group reacts with the sulphhydryl groups of glutathione (GSH) and of proteins. In particular MDA adducts with serum albumin (HSA) and haemoglobin can be used for biological monitoring (7, 56).

For the reasons mentioned in section 4.1, published investigations on the correlation between defined Hb-adducts and the level of external exposure to MDA are not yet available.

4.3 DNA Adducts

In DNA, metabolites of MDA react mostly with the C8 position in desoxyguanosine. *N*-(Desoxyguanosine-8-yl)-MDA is formed in the reaction with *N*-hydroxy-MDA (24).

Biological monitoring via the MDA-DNA adducts in leucocytes is possible in principle (4a). To what extent the detection of DNA adducts can be correlated with a DNA effect in other tissues or with the genotoxicity of MDA requires further investigation. A genotoxic effect can in certain cases also be determined non-specifically by investigating the incidence of DNA strand breaks in leucocytes (41).

5 Methodology

Analysis of MDA in urine or of HSA or Hb adducts takes place by high-pressure liquid chromatography with an LC-UV detector (1, 6, 14, 17, 33, 35, 46, 53) or an LC-ECD detector (30, 48, 49, 50) or, after derivatisation, by capillary gas chromatography with a GC-ECD detector or by GC-MS (2, 8, 9, 10, 38). DNA adducts in the leucocytes are analysed, after being cleaved hydrolytically, by GC-MS (6, 43, 54, 55) or directly using the ^{32}P-postlabelling technique (4a, 45). Using the alkaline elution method, the incidence of DNA strand breaks in the leucocytes can be determined (41).

5.1 Calculation of the MDA Content of Haemoglobin (Hb) and DNA Adducts

During preparation the erythrocytes and leucocytes are isolated from whole blood, washed and then suspended in a volume of physiological saline corresponding to the original volume of blood.

The MDA content of the Hb adduct (ng MDA_{Hb}) and the DNA adducts of the leucocytes (ng MDA_{DNA}), is calculated, allowing for possible preparation losses, by use of the packed cell volumes (PCV), relating the PCV of the erythrocyte or leucocyte suspension to that of the whole blood, according to the equation:

$$C_{\text{Hb or DNA}}(\text{MDA/l blood}) = \frac{PCV_{\text{Blood}}}{PCV_{\text{Suspension}}} \times C_{\text{Suspension}}\ (\text{MDA/l Suspension})$$

$C_{\text{Hb or DNA}}$ = Amount of MDA bound to haemoglobin (erythrocytes) or to DNA (leucocytes) per blood volume in ng MDA per l blood

$C_{\text{Suspension}}$ = MDA level measured in the erythrocyte or leucocyte suspension in ng MDA per l suspension

PCV_{Blood} = Packed cell volume of the whole blood

$PCV_{\text{Suspension}}$ = Packed cell volume of the erythrocyte or leucocyte suspension

6 Evaluation of the Biological Exposure Equivalents for Carcinogenic Substances (EKA)

The current concentration of MDA intermediates after inhalation and percutaneous MDA absorption cannot be directly determined in the human organism, in particular in the liver or in bladder tissue. Conclusions as to individual biologically effective MDA exposure in these target tissues can be deduced, however, from the level of MDA conjugates in erythrocytes (Hb) and also in leucocytes (DNA).

6.1 MDA in Urine

Despite many years of monitoring MDA excretion in urine after industrial exposure to MDA, detailed investigations of its dependence on the inhaled MDA concentration are not available in the literature. Only the British Health and Safety Executive (HSE) recommends in its guidance note on MDA a so-called "yardstick value" of 88 µg MDA/g creatinine which surprisingly was correlated with an 8-hour exposure to only 0.01 ppm (80 µg/m^3) MDA (23). In German occupational-medical practice the value of 100 µg MDA/l urine named earlier in the specific effect definitions has been used to monitor occupational exposure to MDA (29, 47). This value was, however, not scientifically founded. It therefore corresponds to the HSE recommendation. It could not be clarified to what extent the two reference value recommendations take into consideration the percutaneous absorption characteristic of exposure to MDA.

6.2 MDA in Blood Cells

In the literature, adduct data are only described from a few sub-areas, e.g. from isolated Hb samples from persons involved in accidents. After low MDA exposures no significant correlation between the MDA values in the exposure air and those in the adducts in the erythrocytes could be observed (54). The Commission, however, has access to extensive data from chemical plants. The most important results are quoted below. Using current analytical methods no MDA-Hb conjugates are found in persons not exposed to MDA (21, 30).

The typical forms of industrial exposure to MDA are listed in table 2.

Tab. 2: Results of biological monitoring after industrial contact with MDA and various forms of exposure (average values, J. Lewalter, unpublished investigations, 1989–1992)

Air µg MDA/m³	Urine ng MDA/l	Hb adducts ng MDA/l blood	Persons N
1. Continual exposure (routine operation):			
10	270	10	24
20	780	30	12
33	2500	90	6
50	20000	850	3
			→45 persons
2. Intermittent exposure			
	≤50	10	15
	≤100	40	38
	≤1000	170	74
	≤10000	500	137
	≤50000	2300	21
	≤100000	9500	8
			→293 persons
3. Single exposure (intoxication)			
	≤1000	≤50	5
	≤10000	200	29
	≤50000	700	68
	≤100000	2000	43
			→145 persons

Among the persons investigated intermittent contact with MDA with occasional, sometimes very high-level peak exposures, dominated. This can lead to a significant accumulation of the MDA-Hb adducts.

6.3 Conclusions

After occupational contact with MDA concentrations below the currently valid TRK value (100 µg MDA/m³) there is a dose-dependent increase of MDA in urine and the erythrocytic Hb adduct. The data available so far are incomplete. There appear to be difficulties in evaluation due to the high levels of skin absorption. The data in table 2 show a good correlation between the MDA excretion in urine and the Hb adducts in erythrocytes. This indicates the validity of the methods used and their fundamental usefulness for biological monitoring. On the other hand there is no linear correlation between the air values and the biological values. Apparently at workplaces with higher MDA air exposure there is an increased opportunity for immediate skin contact. In table 3 pairs of values are listed which have been assembled, with all necessary reservations, from the values in table 2.

Tab. 3: Concentrations of MDA in air and the associated biological parameters (data taken from table 2)

Air µg MDA/m^3	Urine ng MDA/l	Hb adducts ng MDA/l blood
–	50	10
–	100	40
10	270	–
20	780	–
–	1000	170
33	2500	–
–	10000	500
50	20000	–
–	50000	2300
–	100000	9500

Due to the extreme non-linearity of the relationship between the air concentrations and the biological values in table 3 the DFG Commission for the Investigation of Health Hazards of Chemical Compounds in the Work Area was not prepared to set EKA values for MDA at present. In biological monitoring practice the values in tables 2 and 3 can serve for orientation. MDA will be listed for the time being under the rubric IX/2 (until 1992 VIII/2) of the List of MAK and BAT Values until there are further data or concepts for evaluation.

7 Interpretation of the Data

The MDA excretion determined in urine is not of equal value to the cellular effect of MDA measured in Hb adducts. The cellular effects are the result of individual, metabolic activation determined among other things by the individual acetylation (detoxification) potential and N-oxidation (toxification) potential of the organism. These many and various levels of performance of individual metabolism can only be evaluated at present in the MDA adduct formation with haemoglobin as a marker for reactions with other macromolecules.

In practice, e.g. during occupational medical health surveillance, the routine determination of erythrocytic MDA-Hb adducts is to be recommended. In this way not only the current exposure but also MDA exposures during the previous three months are determined.

Investigations in erythrocytes are not bound to any particular sampling time. On the other hand urine investigations only yield reliable results if sampling takes place immediately after exposure.

8 References

1. Avery, M. J.: Determination of aromatic amines in urine and serum. J. Chromatogr. *488* (1989), 470–475
2. Bailey, E., Brooks, A. G., Bird, I., Farmer, P. B., Street, B.: Monitoring exposure to 4,4′-methylenedianiline by the gas chromatography-mass spectrometry determination of adducts to hemoglobin. Anal. Biochem. *190* (1990), 175–181
3. Bastien, P. G.: Medic: Occupational hepatitis caused by methylenedianiline. Med. J. Australia (1984) 533–535
4. Beard, R. R., Noe, J. T.: Aromatic nitro and amino compounds. In: Patty, F. A. (ed.): Industrial Hygiene and Toxicology, 3rd ed., Vol. 2A. Wiley and Sons, New York 1981, pp. 2413–2489

4a. Beach, A. C., Gupta, R. C.: Human biomonitoring and the ^{32}P-postlabelling assay. Carcinogenesis *13* (7) (1992), 1053–1074

5. Kusters, E.: Biological monitoring of MDA (a letter). Br. Ind. Med. *49* (1992), 72
6. Brunmark, P., Persson, P., Skarping, G.: Determination of 4,4′-methylenedianiline in hydrolysed human urine by micro liquid chromatography with ultraviolet detection. J. Chromatogr. *7* (1992), 1–5
7. Bryant, M. S., Osterman-Golkar, S. M.: Hemoglobin adducts as dosimeters of exposure to DNA-reactive chemicals. CIIT Activities *11* (1991), 1–12
8. Cocker, J., Gristwood, W., Wilson, H. K.: Assessment of occupational exposure to 4,4′-diaminodiphenylmethane (methylenedianiline) by gaschromatography-mass spectrometry analysis of urine. Br. J. Ind. Med. *43* (1986), 620–625
9. Cocker, J., Brown, L. C., Wilson, H. K.: A GC/MS method for the determination of 4,4′-diaminodiphenylmethane and substituted analogues in urine. J. Anal. Toxicol. *12* (1988), 9–14
10. Cocker, J., Boobis, A. R., Davies, D. S.: Determination of the N-acetyl-metabolites of 4,4′-methylenedianiline and 4,4′-methylene-bis-(2-chloro-aniline) in urine. Biomed. Env. Mass Spectro. *17* (1988), 161–167
11. Cohen, S. R.: Yellow staining caused by 4,4′-methylenedianiline exposure. Arch. Dermatol. *121* (1985), 1022–1027
12. Darby, T. D., Johnson, H. J., Northup, S. J.: An evaluation of a polyurethane for use as a medical grade plastic. Toxicol. Appl. Pharmacol. *46* (1978), 449–453
13. Deichmann, B., MacDonald, W. E., Coplan, M., Woods, F., Blum, E.: Di-(4-aminophenyl)-methane(MDA): 4-year dose feeding study. Toxicology *11* (1978), 185–188
14. Ernes, D. A., Hanshumaker, D. T.: Determination of extractable methylene-bis(aniline) in polyurethane films by liquid chromatography. Anal. Chem. *55* (1983), 408–409
15. International Agency for Research on Cancer (IARC): 4,4′-Methylenedianiline and its dihydrochloride. In: The evaluation of the carcinogenic risk of chemicals to humans. IARC Monograph No. 39. IARC, Lyon 1986, pp. 347–365
16. International Agency for Research on Cancer (IARC): 4,4′-Methylenediphenyl diisocyanate and polymethylene polyphenyl isocyanate. In: The evaluation of the carcinogenic risk of chemicals to humans. IARC Monograph No. 19. IARC, Lyon 1974, pp. 314 ff.
17. Frei, R. W., Nielen, M. W. F., Brinkman, U. A. T. H.: Handling of Environmental and biological samples via precolumn technologies. Inter. J. Environ. Chem. *25* (1986), 3–35
18. Fukushima, S., Shibata, M., Hibino, T., Yoshimura, T., Hirose, M., Ito, N.: Intrahepatic bile duct proliferation induced by 4,4′-diaminodiphenylmethane in rats. Toxicol. Appl. Pharmacol. *48* (1979), 145–155
19. Grayson, M. (ed.): Methylenedianiline. Kirk-Othmer-Encyclopedia of chemical technology, Vol. 2. Wiley and Sons, New York 1978, pp. 338 ff.
20. Henderson, R. F., Bechtold, W. E., Bond, J. A., Sun, J. D.: The Use of Biological Markers in Toxicology. Toxicology *20* (1989), 65–82
21. Schütze, D., Sepai, D., Lewalter, J., Miske, L., Henschler, D., Sabbioni, G.: Biomonitoring of workers exposed to 4,4′-methylenedianiline (MDA) or 4,4′-methylenediphenyldiisocyanate (MDA). Carcinogenesis, (1995) in press
22. Henschler, D. (ed.): Gesundheitsschädliche Arbeitsstoffe. Toxikologisch-arbeitsmedizinische Begründung von MAK-Werten: 4,4′-Diaminodiphenylmethan und -Dihydrochlorid. 12. Lfg. VCH-Verlagsgesellschaft, Weinheim 1987

23. Fairhurst, S., South, D., Williams, C., Cocker, J., Brown, R.: 4,4′-Methylenedianiline. Criteria document for an occupational exposure limit. Health and Safety Executive, HMSO, London 1993
24. Kadlubar, F. F., Miller, J. A., Miller, E. C.: Hepatic microsomal N-glucuronidation and nucleic acid binding of N-hydroxyarylamines in relation to urinary bladder carcinogenesis. Cancer Res. *37* (1977), 805–814
25. Koppelmann, H., Scheuer, P. J., Williams, R.: The liver lesion of the Epping jaundice. Quart. J. Med. *35* (1966), 553–564
26. Koppelmann, H., Robertson, M. H., Sanders, P. G., Ash, I.: The Epping jaundice. Br. Med. J. *1* (1966), 514–516
27. Larroque, M., Vian, L., Blaise, A., Brun, S.: Méthodes de dosage des monomères résiduels des résines époxydiques dans des simultans du vin. J. Chromatogr. *445* (1988), 107–117
28. LeVine, M. J.: Occupational photosensitivity to diaminodiphenylmethane. Cont. Derma. *9* (1983), 488–490
29. Lewalter, J.: Auswertung der arbeitsmedizinischen Vorsorge- und Biomonitoringuntersuchungen von MDA-Belasteten (1980 bis 1992). III MDA review meeting, Straßburg 1993
30. Lewalter, J., Bussmann, R.: Simultanbestimmung von Aminoaromaten in Harn, Plasma und Hämoglobin mit HPLC-ECD. In: Henschler, D. (ed.): Analytische Methoden zur Prüfung gesundheitsschädlicher Arbeitsstoffe. Bd. 2: Analysen in biologischem Material, 11. Lfg., Deutsche Forschungsgemeinschaft, VCH Verlagsgesellschaft, Weinheim 1993
31. McCarthy, D. J., Struck, R. F., Shih, T.-W., Suling, W. J., Hill, D., Enke, S. E.: Disposition and metabolism of the carcinogen reduced Michler's ketone in rats. Cancer Res. *42* (1982), 3475–3479
32. McGill, D. B., Motto, J. D.: An industrial outbreak of toxic hepatitis due to methylene-dianiline. N. Engl. J. Med. *291* (1974), 278–282
33. National Institute for Occupational Safety and Health (NIOSH): 4,4′-methylene-bis-(phenylisocyanate) (MDI). NIOSH Manual of Analytical Methods, Vol. 7, Method No. 347. U. S. Government Printing Office, Washington DC. 1981, pp. 1–13
34. National Institute for Occupational Safety and Health (NIOSH): NIOSH Health hazard evaluation report Boeing Verol Co., HETA 82-146-1388, Philadelphia, PA. Cincinnati, OH., 1983
35. Neumeister, C.E.: Analysis of urine to monitor exposures to benzidine, *o*-Dianisidine, *o*-Toluidine and 4,4′-Methylenedianiline. Appl. Occup. Environ. Hyg. *6* (1991), 953–958
36. Parodi, S., Taningher, M., Russo, P., Pala, M., Tamaro, M., Monti-Bragadin, C.: DNA-damaging activity in vivo and bacterial mutagenicity of sixteen aromatic amines and azo-derivatives, as related quantitatively to their carcinogenicity. Carcinogenicity *2* (1981), 1317–1326
37. Pludre, G., Karlowski, K., Mankowska, M., Woggon, H., Uhde, W.-J.: Toxicological and chemical studies of some epoxy resins and hardeners. I.: Determination of acute and subacute toxicity of phthalic acid anhydride, 4,4′-diaminodiphenylmethane and of the epoxy resin: Epilox EG-34. Acta. Pol. Pharm. *26* (1969), 352–364
38. Rosenberg, C., Savolainen, H.: Mass fragmentographic determination of urinary amine metabolites in rats exposed to degradation products from heated rigid polyurethane. J. Chromatogr. *358* (1986), 385–392
39. Roy, C. W., McSorley, P. D., Syme, I. G.: Methylene dianiline: a new toxic cause of visual failure with hepatitis. Hum. Toxic. *4* (1985), 61–66
40. Seldén, A., Berg, P., Jakobsson, R., Lavel de, J.: Methylene dianiline: assessment of exposure and cancer morbidity in power generator workers. Int. Arch. Occup. Environ. Health *63* (1992), 403–408
41. Skare, N., Schretel, K.R.: Alkaline elution of rat testicular DNA: Detection of DNA strand breaks after in vivo treatment with chemical mutagens. Mutat. Res. *130* (1984), 283–294
42. Steinhoff, D., Grundmann, E.: Zur kanzerogenen Wirkung von 4,4′-Diaminodiphenylmethan und 2,4-Diaminodiphenylmethan. Naturwissenschaften *57* (1970), 247–248
43. Tamura, N., King, Ch.M.: Comparative survival of aminobiphenyl- and aminofluorene-substituted plasmid DNA in *Escherichia coli* Uvrendonuclease deficient strains. Carcinogenesis *11* (1990), 535–540
44. Tanaka, K., Ino, T., Sawahata, T., Marui, S., Igaki, H., Yashima, H.: Mutagenicity of N-acetyl and N,N′-diacetyl derivatives of 3 aromatic amines used as epoxy-resin hardeners. Mutat. Res. *143* (1985), 11–15
45. Talaska, G., Dooley, K.L., Kadlubar, F.F.: Detection and characterization of carcinogen-DNA adducts in exfoliated urothelial cells from 4-aminobiphenyl-treated dogs by 32P-postlabelling and subsequent thin-layer and high-pressure liquid chromatography. Carcinogenesis *11* (1990), 639–646

46. Tiljander, A., Skarping, G.: Determination of 4,4´-methylenedianiline in hydrolysed human urine using liquid chromatography with UV detection and peak identification by absorbance ratio. J. Chromatogr. *511* (1990), 185–194
47. Spezifische Einwirkungsdefinitionen, ZH1/600. Hauptverband der gewerblichen Berufsgenossenschaften, Bonn 1979, pp. 75–78
48. Tiljander, A.: Chromatographic determination of 4,4´-methylenedianiline in biological fluids. Dissertation, Department of Occupational and Environmental Medicine, Lund 1989
49. Tortoreto, M., Catalani, P., Bianchi, M., Blonda, C., Pantarotta, C., Paglialunga, S.: Determination of 4,4´-diaminodiphenylmethane in blood by gasliquid chromatography with electron-capture detection. J. Chromatogr. *262* (1983), 367–372
50. Trippel-Schulte, P., Zeiske, J., Kettrup, A.: Trace analysis of selected benzidine and diaminodiphenylmethane derivates in urine by means of liquid chromatography using precolumn sample preconcentration, UV and electrochemical detection. Chromatogr. *22* (1986), 138–146
51. Vaudaine, Lery, N., Diter, J. N., Droin, M., Chamaillard, C.: Le diamino diphenyl methane: un example de toxicolovigilance dans les ateliers de fabrication de rhone-poulenc-industrie. Journal de Toxicologie Medicale *2* (1982), 207–214
52. Weisburger, E. K., Murthy, A. S. K., Lilja, H. S., Lamb, J. C.: IV: Neoplastic response of F344 rats and B6C3F1 mice to the polymer and dyestuff intermediates 4,4´-methylene bis-(N,N-dimethyl)-benzeneamine, 4,4´-oxydianiline, and 4,4´-methylenedianiline. J. Natl. Cancer Inst. *72* (1984), 1457–1463
53. Winkler, E., Hohaus, E., Felber, E.: High-performance liquid chromatographic separation of ethylenediamine and triethylenetetramine after derivatization with salicylaldehyde-diphenylboron chelate and separation of 4,4´-diaminodiphenylmethane by high-performance liquid chromatography. J. Chromatogr. *436* (1988), 447–454
54. Schütze, D.: Methoden und Grundlagen zum Expositionsnachweis und zur Risikoermittlung von Bis-(4-Aminophenyl)-methan und davon abgeleiteten Industriechemikalien. Dissertation, Universität Würzburg (1993)
55. Bakthavachalam, J., Abdel-Baky, S., Giese, R. W.: Release of 2-aminofluorene from N-(desoxyguanosin-8-yl)-2-aminofluorene by hydrazinolysis. J. Chromatogr. *538* (1991), 447–451
56. Neumann, H. G., Birner, G., Kowallik, P., Schütze, D., Zwirner-Baier, I.: Hemoglobin adducts of N-substituted aryl compounds in exposure control and risk assessment. Environ. Health Perspect. *99* (1993), 65–69

Author: J. Lewalter
Approved by the Working Group: 27th May 1993

2-Naphthylamine

Synonyms	2-Aminonaphthalene β-Naphthylamine
Formula	$C_{10}H_9N$ 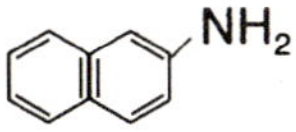
Molecular weight	143.2
Melting point	111 °C
Boiling point	306 °C
MAK [last established: 1970]	Carcinogenic substance: category IIIA1 of the List of MAK and BAT Values

1 Metabolism and Kinetics

2-Naphthylamine is absorbed rapidly via inhalation and especially percutaneously and then excreted via the kidneys in the form of the free or conjugated amine and conjugated metabolites with a half-time of up to seven hours (40, 43).

Some of the 2-naphthylamine is oxidised in the liver by cytochrome P450 enzymes to *N*-hydroxy-2-naphthylamine and then partly glucuronidised. Some of the free *N*-hydroxy-2-naphthylamine can be oxidised in erythrocytes to 2-nitrosonaphthalene with simultaneous formation of methaemoglobin. The reactive nitroso compound can react with glutathione ($t_{1/2} < 7$ hours) and globin ($t_{1/2} \gg 10$ days) to sulfinamides (5, 8, 9, 12, 15, 27, 34, 38, 39 ,41, 42).

For further details of the metabolism of 2-naphthylamine the occupational-medical toxicological documentation for BAT values for 4-aminodiphenyl can be referred to (22). The general metabolic steps are shown in figure 1.

In contrast to *N*-hydroxy-1-naphthylamine, in the case of *N*-hydroxy-2-naphthylamine the so-called "1,4-Bamberger rearrangement" to 2,5-aminonaphthol is inhibited so that the stable metabolite *N*-hydroxy-2-naphthylamine released from the glucuronic acid conjugate in the acidic urine can react after oxidation to 2-nitrosonaphthalene with macromolecules of the bladder mucous membrane (40). This explains the tumour-initiating effect of 2-naphthylamine on the cells of the epithelium of the urinary tract. The individual disposition towards the carcinogenic effect of 2-naphthylamine depends to a great extent on the individual status as a slow or fast acetylator (see section 1.1).

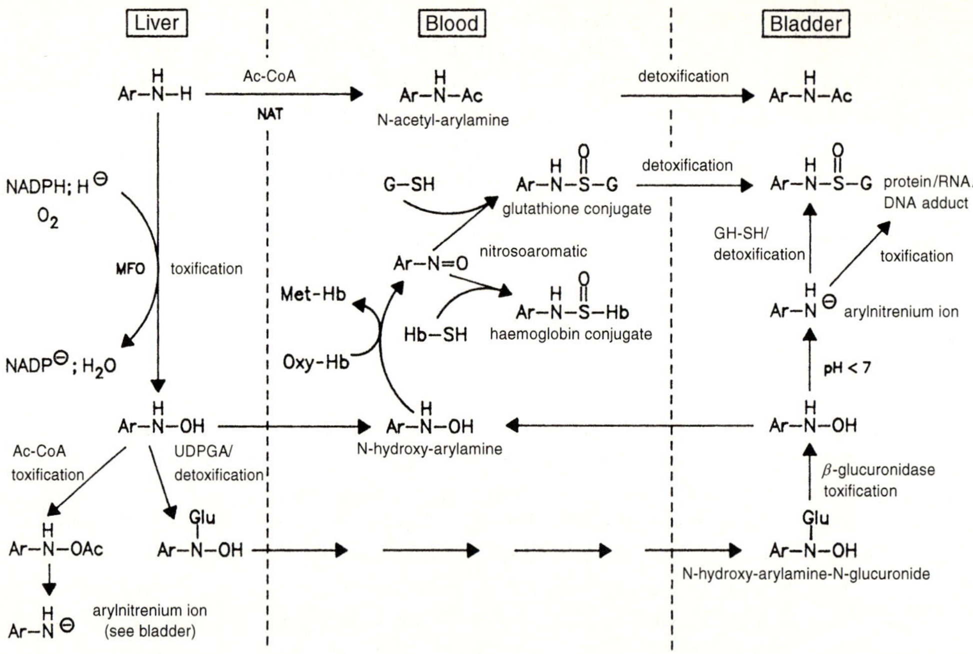

MFO: mixed function oxidases
G-SH: glutathione
Hb-SH: haemoglobin
Ar-: aromatic ring

Glu: glucuronic acid
Ac: acetyl residue
Ac-CoA: acetyl co-enzyme A
UDPGA: uridine diphosphate-glucuronic acid
NAT: *N*-acetyltransferase

Fig. 1: Metabolism of primary aminoaromatics: formation and transport of the detoxification stages in the various compartments (40)

Variability of 2-Naphthylamine Effect

Correlations between protein and DNA binding of amino aromatics repeatedly observed in animal experiments (17, 30, 31, 32, 37) may indicate the correctness of individual protection by appropriate monitoring of the protein(Hb) dose but do not allow evaluation of any associated risk in the individual case (7, 46). Adduct formation is apparently largely avoided if the incorporated 2-naphthylamine is completely acetylated (and thus eliminated rapidly via urine) (19, 20). In this respect the acetylator status has an important influence on the amount of adduct formation, which is particularly marked in "slow acetylators" (19, 28).

Relationships between acetylator phenotype and the occurrence of bladder carcinoma were described for the first time in 1979 (24, 25, 29). In epidemiological studies it was found that fast acetylators are represented markedly less often among patients with tumours of the efferent urinary passages than slow acetylators despite comparable exposure to carcinogenic amino aromatics (10, 13, 24, 25, 26, 28, 29). Hansen et al. found 61.9 % slow acetylators in 105 patients with bladder tumours compared to only 42.9 % in patients without bladder carcinoma (13). The ratio was as large as 70:30 for occupational bladder tumours. Observations in collectives with complete occupational anamneses show even more significant ratios (90 : 10) (21). Not only the amount of 2-naphthylamine taken up, but also the individual metabolic disposition decides the kind and extent of the individually formed 2-naphthylamine-protein and 2-naphthylamine-DNA conjugates and thus possible genotoxic effects (1, 7, 31, 46). In general after acute 2-naphthylamine intoxication conspicuous levels of 2-naphthylamine-haemoglobin conjugates are observed only in weak acetylators (20a).

2 Critical Toxicity

The currently available data has shown 2-naphthylamine to be carcinogenic in humans (16). Sessile or pediculate papillomas which become malignant are characteristic. Also primary transitional cell carcinomas have been observed.

In contrast to the biological monitoring of aniline for which the BAT value is oriented to the Met-Hb level (14) biological monitoring of persons exposed to 2-naphthylamine must be aimed at minimising or avoiding (additional) genotoxic risks (38a). A basis for the prophylaxis of persons exposed to 2-naphthylamine can therefore only be gained from the level of environmental, at present unavoidable 2-naphthylamine background exposure, which is found in the general population today e.g. as a result of smoking.

3 Exposure and Effects

In the case of 2-naphthylamine every exposure which can lead to inhalation or percutaneous absorption is to be regarded as external exposure. The internal exposure consists of the biologically relevant concentration of the reactive metabolites in the target organ, above all in the efferent urinary passages. Parameters of internal exposure are the 2-naphthylamine concentrations in body fluids such as blood, plasma and urine which correlate with the internal exposure under the specific pharmacokinetic conditions. 2-Naphthylamine adducts in erythrocytes (Hb adducts) or leucocytes (DNA adducts) are understood as parameters of 2-naphthylamine effects.

Monitoring the incidence of DNA strand breaks in leucocytes of persons exposed to 2-naphthylamine provides a further, albeit non-specific parameter of effect which is suitable for evaluation of acute 2-naphthylamine intoxication.

4 Selection of the Indicators

The internal 2-naphthylamine exposure and effects possible after contact with 2-naphthylamine can be determined via analysis of the substance in urine, blood and plasma, as well as via protein and DNA adducts in the blood cells.

4.1 2-Naphthylamine in Urine

2-Naphthylamine exposure can be determined via the excretion of free and conjugated 2-naphthylamine in urine (23, 33). Despite comprehensive investigations an exact correlation between inhaled 2-naphthylamine concentrations and the urinary excretion of 2-naphthylamine is not yet available. This is a result of the fact that during exposure to 2-naphthylamine the dominant exposure route is often percutaneous and its share of the total body burden from 2-naphthylamine cannot be quantitatively determined. In occupational-medical practice 2-naphthylamine excretion after contact with 2-naphthylamine is often evaluated on the basis of so-called "empirical values".

4.2 2-Naphthylamine Protein Adducts

According to the investigations of various research groups, the product of metabolism of 2-naphthylamine, 2-nitrosonaphthalene, reacts with the sulphydryl groups of glutathione (GSH) (6) and proteins (4). In particular 2-naphthylamine adducts which are formed with human serum albumin (HSA) and haemoglobin can be used for biological monitoring (4).

For the reasons already mentioned in section 4.1, investigations on the correlation between defined Hb adducts and the extent of external exposure to 2-naphthylamine are, however, lacking.

4.3 DNA Adducts

Within DNA, position C8 in deoxyguanosine is the target for reaction with active metabolites of 2-naphthylamine. *N*-(Deoxyguanosine-8-yl)-2-naphthylamine is formed in the reaction with the metabolite, *N*-hydroxy-2-naphthylamine (17).

Biological monitoring of the 2-naphthylamine adducts in leucocytes is possible in principle. To what extent the detection of DNA adducts can be correlated with an effect on DNA or the genotoxicity of 2-naphthylamine requires further investigation (7, 31, 46). A genotoxic effect can be determined also non-specifically by determining the incidence of DNA strand breaks (2).

5 Methodology

Analysis of 2-naphthylamine in urine takes place directly, and in Hb adducts after cleavage by high-pressure liquid chromatography with an LC-ECD detector or, after derivatisation, by capillary gas chromatography with an GC-ECD detector or using GC/MS. HSA adducts can be screened qualitatively by isoelectric focusing or determined quantitatively by LC, GC or GC/MS after cleavage of the 2-naphthylamine. The 2-naphthylamine in the DNA adducts of the leucocytes is analysed after cleavage by GC/MS or directly using the ^{32}P-postlabelling technique. The methods of analysis are described in the literature (4, 15, 18, 45). The incidence of DNA strand breaks in the leucocytes can be checked using the alkaline elution (44).

Cytological examination of the urinary sediment with the Papanicolaou test (35) should form part of the clinical prophylaxis against amino aromatics. After repeated microhaematuria or the detection of pathological cells in the urinary sediment, a urological examination with a cystoscope should be carried out (11).

Calculation of the 2-Naphthylamine Content of Haemoglobin (Hb) or DNA Adducts

During preparation the erythrocytes and leucocytes are isolated from whole blood, washed and then suspended in a volume of physiological saline corresponding to the original volume of blood.

The 2-naphthylamine content of the Hb adduct (µg 2-NA_{Hb}) or of the DNA adducts in the leucocytes (µg 2-NA_{DNA}) can be calculated allowing for possible preparation losses by use of the packed cell volumes (PCV) relating the PCV of the erythrocyte or leucocyte suspension to that of the whole blood, as follows:

$$C_{\text{Hb or DNA}}(\text{2-NA/l blood}) = \frac{PCV_{\text{Blood}}}{PCV_{\text{Suspension}}} \times C_{\text{Suspension}}\ (\text{2-NA/l suspension})$$

$C_{\text{Hb or DNA}}$	= Content of 2-NA bound to haemoglobin (erythrocytes) or to DNA (leucocytes) per blood volume in µg 2-NA per l blood
$C_{\text{Suspension}}$	= 2-NA level measured in the erythrocyte or leucocyte suspension in µg 2-NA per l suspension
PCV_{Blood}	= Packed cell volume of whole blood
$PCV_{\text{Suspension}}$	= "Packed cell volume" of the erythrocyte or leucocyte suspension

6 Currently Available Data on Biological Monitoring Parameters

The concentration of 2-naphthylamine intermediates occurring at a given time after respiratory and percutaneous 2-naphthylamine absorption cannot be directly determined in human bladder tissue. Conclusions as to the biologically effective 2-naphthylamine exposure of the individual can, however, be derived from the level of 2-naphthylamine conjugated in erythrocytes and leucocytes.

6.1 2-Naphthylamine in Urine

Investigations on the dependence of 2-naphthylamine excretion in urine on the inhaled 2-naphthylamine concentration are lacking in the literature. In particular at industrial workplaces percutaneous uptake of 2-naphthylamine dominates. After contact with amino aromatics, detectable urinary excretion of 2-naphthylamine can most often be explained by percutaneous absorption.

The available database does not allow an evaluation of EKA values for renal excretion of 2-naphthylamine. In practice, occupational-medical monitoring of exposure to 2-naphthylamine has made use of empirical "2-naphthylamine monitoring values" of the order of magnitude of 1.0 µg 2-naphthylamine/g creatinine, however without scientific justification (20a).

6.2 2-Naphthylamine in Blood Cells

In non-smokers not exposed to 2-naphthylamine, 2-naphthylamine-Hb conjugates of on average 2 ng amino aromatics per l blood (range: 1–6) were found (4, 11, 20a, 47). Smoking approx. 20 cigarettes per day increases the level of 2-naphthylamine-Hb adduct to approx. 5 ng amino aromatics per l blood (range: 2–17) (4, 20a) and leads to total urinary excretion of approx. 30 ng 2-naphthylamine per g creatinine (range: 2-380) (Tab. 1). The 2-naphthylamine dose inhaled from main stream smoke amounts to approx. 34 ng 2-naphthylamine (daily total per person). In addition, the side stream smoke contains approx. 1340 ng 2-naphthylamine (also as daily total) (36, 47). The DNA adduct level in the leucocytes was also tested and was below the analytical detection limit of 1.0 µg 2-naphthylamine/l blood. In the leucocytes the incidence of DNA strand breaks was not increased (20a).

Tab. 1: Influence of smoking and acetylator status on amino aromatic conjugates in haemoglobin (20a)

			Level of Hb conjugates		
Smoking status	Persons n	N-Acetyl-transferase activity	4-Amino-diphenyl	2-Naphthyl-amine	Benzidine
Smoker approx. 1 pack = 19 cigarettes per	25	low	21 ± 3.5	6 ± 1.0	≈1
24 hours	23	high	17 ± 3.0	4 ± 1.5	≈1
Non-smoker	17	low	6 ± 1.0	2 ± 1.0	<1
	15	high	4 ± 1.0	1 ± 0.5	<1

6.3 Conclusions

As a carcinogenic substance, 2-naphthylamine is listed in section IX,2 in the list of MAK and BAT values with carcinogenic substances for which correlations (exposure equivalents of carcinogenic substances, EKA) cannot be evaluated, or only incompletely evaluated. The currently available information is therefore presented here as documentation. According to the most detailed data available to the Commission, the biological parameters for the evaluation of 2-naphthylamine exposure cannot be correlated with any quantifiable external exposure, but are correlated with the occurrence of Hb and DNA adducts. According to these data 2-naphthylamine exposures with

urinary excretion levels below 1 μg 2-naphthylamine/g creatinine and 2-naphthylamine-Hb adduct levels below 0.1 μg 2-NA/ blood

do not yet lead to DNA adduct levels in leucocytes detectable using currently available methods of analysis. In practice, biological monitoring of persons exposed to 2-naphthylamine can be based on this as long as the setting of EKA values is not possible.

7 References

1. Beland, F. A., Kadlubar, F. F.: Formation and persistence of arylamine DNA adducts in vivo. Environ. Health Perspect. *62* (1985), 19–30
2. Birnboim, H. C., Kanabus-Kaminska, M.: The production of DNA strand breaks in human leukocytes by superoxid anion may involve a metabolic process. Proc. Natl. Acad. Sci. USA *82* (1985), 6820–6824
3. Bryant, M. S., Skipper, P. L., Tannenbaum, S. R., Maclure, M.: Hemoglobin adducts of 4-aminodiphenyl in smokers and non-smokers. Cancer Res. *47* (1987), 602–608
4. Bryant, M. S., Vineis, P., Skipper, P. L., Tannenbaum, R.: Hemoglobin adducts of aromatic amines: Association with smoking status and type of tobacco. Proc. Natl. Acad. Sci. USA *85* (1988), 9788–9791
5. Eyer, P., Lierheimer, E.: Biotransformation of nitrosobenzene in the red cell and the role of glutathione. Xenobiotica *10* (1980), 517–526
6. Eyer, P.: The red cell as a sensitive target for activated toxic arylamines. Arch. Toxicol. Suppl. *6* (1983), 3–12
7. Farmer, P. B., Neumann, H. G., Henschler, D.: Estimation of exposure of man to substances reacting covalently with macromolecules. Arch. Toxicol. *60* (1987), 251–260
8. Flammang, T. J., Kadlubar, F. F.: Acetyl-CoA-dependent, cytosol-catalyzed binding of carcinogenic N-hydroxy-arylamines to DNA. In: Boobies, A. R., Caldwell, J., DeMatteis, F., Elcombe, C. R. (eds.): Microsomes and Drug Oxidations, Taylor and Francis, London 1985, pp. 190–197
9. Frederick, C. B., Hammons, G. J., Beland, F. A., Yamazoe, Y., Güngerich, F. P., Zenser, T. V., Ziegler, D. M., Kadlubar, F.: N-oxidation of primary aromatic amines in relation to chemical carcinogenesis. In: Gorrod, J. W., Damani, L. A. (eds.): Biological oxidation of nitrogen in organic molecules. Ellis Horwood, Chichester *13* (1985), 131–148
10. Glowinski, J. B., Radtke, H. E., Weber, W. W.: Genetic variation in N-acetylation of carcinogenic arylamines by human and rabbit liver. Molecular Pharmacology *14* (1978), 940–949
11. Hauptverband der gewerblichen Berufsgenossenschaften (ed.): Grundsatz G33: Aromatische Nitro- und Aminoverbindungen, Berufsgenossenschaftlicher Grundsatz für arbeitsmedizinische Vorsorgeuntersuchungen, II.A., Gentner, Stuttgart 1981
12. Hanna, P. E., Banks, R. B.: Arylhydroxylamines and arylhydroxamic acids: conjugation reactions. In: Anders, M. W. (ed.): Bioactivation of foreign Compounds. Academic Press, Orlando 1985, pp. 375–402

13. Hanssen, H. P., Agarwal, D. P., Gödde, H. W., Bucher, H., Huland, H., Brachmann, W., Ovenbeck, R.: Association of N-acetyltransferase polymorphism and environmental factors with bladder carcinogenesis. Eur. Urol. *11* (1985), 263–266
14. Henschler, D., Lehnert, G. (ed.): Biologische Arbeitsstoff-Toleranz-Werte (BAT-Werte) und Expositionsäquivalente für krebserzeugende Arbeitsstoffe (EKA). Arbeitsmedizinisch-toxikologische Begründungen, 3rd issue. VCH Verlagsgesellschaft, Weinheim 1986
15. Kadlubar, F. F., Miller, J. A., Miller, E. C.: Hepatic microsomal N-glucuronidation and nucleic acid binding of N-hydroxy arylamines in relation to urinary bladder carcinogenesis. Cancer Res. *37* (1977), 805–814
16. Korallus, U., Lewalter, J.: Aromatische Amine. In: Konietzko, J., Dupuis, H. (eds.): Handbuch der Arbeitsmedizin. IV-2.30.2 (1989), 1–77
17. Kriek, E., Westra, J. G.: Metabolic activation of aromatic amines and amides and interactions with nucleic acids. In: Grover, P. L. (ed.): Chemical carcinogens and DNA, Vol. II. CRC Press, Boca Raton 1979, pp. 1–28
18. Levy, G., Weber, W. W.: High-performance liquid chromatographic analysis of ^{32}P-postlabeled DNA-aromatic carcinogen adducts. Analyt. Biochem. *174* (1988), 381–392
19. Lewalter, J., Korallus, U.: Blood proteins conjugates and acetylation of aromatic amines: New findings on biological monitoring. Int. Arch. Occup. Environ. Health *56* (1985), 179–196
20. Lewalter, J.: Neue methodische Ansätze in der biologischen Überwachung Arbeitsstoff-Exponierter. Verh. Dt. Ges. Arbmed., 25. Jahrestagung. Gentner, Stuttgart 1985, pp. 151–166
20a. Lewalter, J.: Auswertung der arbeitsmedizinischen Vorsorge- und Biomonitoring-Untersuchungen von Aminoaromaten-Belasteten, unpublished investigations
21. Lewalter, J., Miksche, L., Kiesselbach, N.: Empfehlungen zur arbeitsmedizinischen Prävention expositions- und dispositionsbedingter Arbeitsstoff-Beanspruchungen. Verh. Dt. Ges. Arbmed., 31. Jahrestagung. Gentner, Stuttgart 1991, pp. 135–139
22. Henschler, D., Lehnert, G. (eds.): Biologische Arbeitsstoff-Toleranz-Werte (BAT-Werte) und Expositionsäquivalente für krebserzeugende Arbeitsstoffe (EKA). Arbeitsmedizinisch-toxikologische Begründungen, 4th issue. VCH Verlagsgesellschaft, Weinheim 1989
23. Linch, A. L.: Biological monitoring for industrial chemical exposure control. CRC Press, Cleveland, Ohio 1974
24. Lower, G. M. jr.: N-Acetyltransferase phenotype and risk in industrial urinary bladder cancer: Approaches to high risk groups. In: Deichmann (ed.): Toxicology and Occupational Medicine. Elsevier, Niederlande 1979, pp. 209–219
25. Lower, G. M. jr., Nilsson, T., Nelson, C. E., Wolf, H., Gamsky, T. E., Bryan, G. T.: N-Acetyltransferase phenotype and risk in urinary bladder cancer: Approaches in molecular epidemiology. Environ. Health Perspect. *29* (1979), 71–79
26. Lower, G. M. jr.: Arylamines and bladder cancer causality. Application of conceptual and operational criteria. Clin. Pharmacol. Ther. *34* (1983), 129–135
27. McQueen, Ch. A., Way, B. M., Williams, G. M.: Biotransformation of aromatic amines to DNA-damaging products by urinary bladder organ cultures. Carcinogenesis *8* (1987), 401–404
28. Mommsen, S., Barford, N. M., Aagaard, J.: N-Acetyltransferase phenotype in the urinary bladder carcinogenesis of a low-risk population. Carcinogenesis *6* (1985), 199–201
29. Mommsen, S., Aagaard, J.: Susceptibility in urinary bladder cancer: Acetyltransferase phenotypes and related risk factors. Cancer Lett. *32* (1986), 199–205
30. Neumann, H. G.: Analysis of hemoglobin as a dose monitor for alkylating and arylating agents. Arch. Toxicol. *56* (1987), 1–6
31. Neumann, H.G.: Concepts for assessing the internal dose of chemicals in vivo. In: Fowler, B. A. (ed.): Mechanism of cell injury: implications for human health. Wiley and Sons, New York 1987, pp. 241–254
32. Neumann, H. G.: Hemoglobin binding in exposure control and risk assessment of aromatic amines. In: International Conference on detection methods for DNA-damaging agents in man: Application in cancer, epidemiology and prevention. Espoo, Finland 1987
33. Nony, C. R., Bowman, M. C.: Carcinogens and analogs: trace analysis of thirteen compounds in admixture in waste water and human urine. Int. J. Env. Anal. Chem. *5* (1978), 203
34. Oglesky, L. A., Flammang, T. J., Tullis, D. L., Kadlubar, F. F.: Rapid absorption, distribution and excretion of carcinogenic N-hydroxyarylamines after direct urethral instillation into the rat urinary bladder. Carcinogenesis *2* (1981), 15–20

35. Papanicolaou, G. N.: Atlas of exfolia tire cytology, Harvard University Press, Cambridge, Mass. 1954
36. Patrianakos, C., Hoffmann, D.: Chemical studies on tobacco smoke. LXIV: On the analysis of aromatic amines in cigarette smoke. J. Analyt. Tox. *3* (1973), ISO–154
37. Pereira, M. A., Lin, L. H. C., Chang, L. W.: Dose-dependency of 2-acetylaminofluorene binding to liver DNA and hemoglobin in mice and rats. Toxicol. Appl. Pharmacol. *60* (1981), 472–478
38. Poupko, J. M., Haern, W. L., Radomski, J. L.: N-Glucuronidation of N-hydroxy aromatic amines. A mechanism for their transport and bladder-specific carcinogenicity. Toxicol. Appl. Pharmcol. *50* (1979), 479–484
38a. Purchase, J. F. H., Kalinowski, A. E., Ishmael, J., Wilson, J., Gore, C. W., Chart, J. S.: Lifetime carcinogenicity study of 1- and 2-naphthyl-amine in dogs. Br. J. Cancer *44* (1981), 892–902
39. Radomski, J. L., Deichmann, W. B., Glass, E. M.: Carcinogenicity testing of N-hydroxy and other oxidation and decomposition products of 1- and 2-naphthylamine. Cancer Res. *31* (1971), 1461–1467
40. Radomski, J. L.: The primary aromatic amines: their biological properties and structure-activity relationships. Ann. Rev. Pharmacol. Toxicol. *19* (1979), 129–157
41. Radomski, J. L., Conzelmann jr., G. L., Rey, A. A., Brill, E.: N-oxidation of certain aromatic amines, acetamides and nitro compounds by monkeys and dogs. J. Natl. Cancer Inst. *50* (1973), 989–995
42. Radomski, J. L., Brill, E.: Bladder cancer induction by aromatic amines: Roles of N-hydroxy metabolites. Science *167* (1979), 992–993
43. Shuker, C. K., Batt, S., Rystedt, I., Berlin, M.: The health effect of aromatic amines. A review, a technical report. Monitoring and Assessment Research Centre, King's College, London 1986
44. Skare, J. A., Schretel, K. R.: Alkaline elution of rat testicular DNA: Detection of DNA strand breaks after in vivo treatment with chemical mutagens. Mutat. Res. *130* (1984), 283–294
45. Stillwell, W. G., Bryant, M. S., Wishnok, J. S.: GC/MS Analysis of biologically important aromatic amines; application to human dosimetry. Biomed. Environ. Mass. Spectrom. *14* (1987), 221–227
46. Swenberg, J. A., Richardson, F. C., Boucheron, J. A., Dyroff, M. C.: Relationships between DNA adduct formation and carcinogenesis. Environ. Health Perspect. *62* (1982), 177–183
47. Tannenbaum, S. R., Bryant, M. S., Skipper, P. L., Maclure, M.: Hemoglobin adducts of tobacco related aromatic amines: Application to molecular epidemiology. Banbury Rep. *23* (1986), 63–75

Author: J. Lewalter
Approved by the Working Group: 19th December 1991